INSIDE WOOD

MASTERPIECE OF NATURE

Published by

THE AMERICAN FORESTRY ASSOCIATION

919 Seventeenth Street, N.W., Washington, D. C. 20006

Frontispiece

Cross Section
Alder Twig Natural Size, and Magnified 40 Times

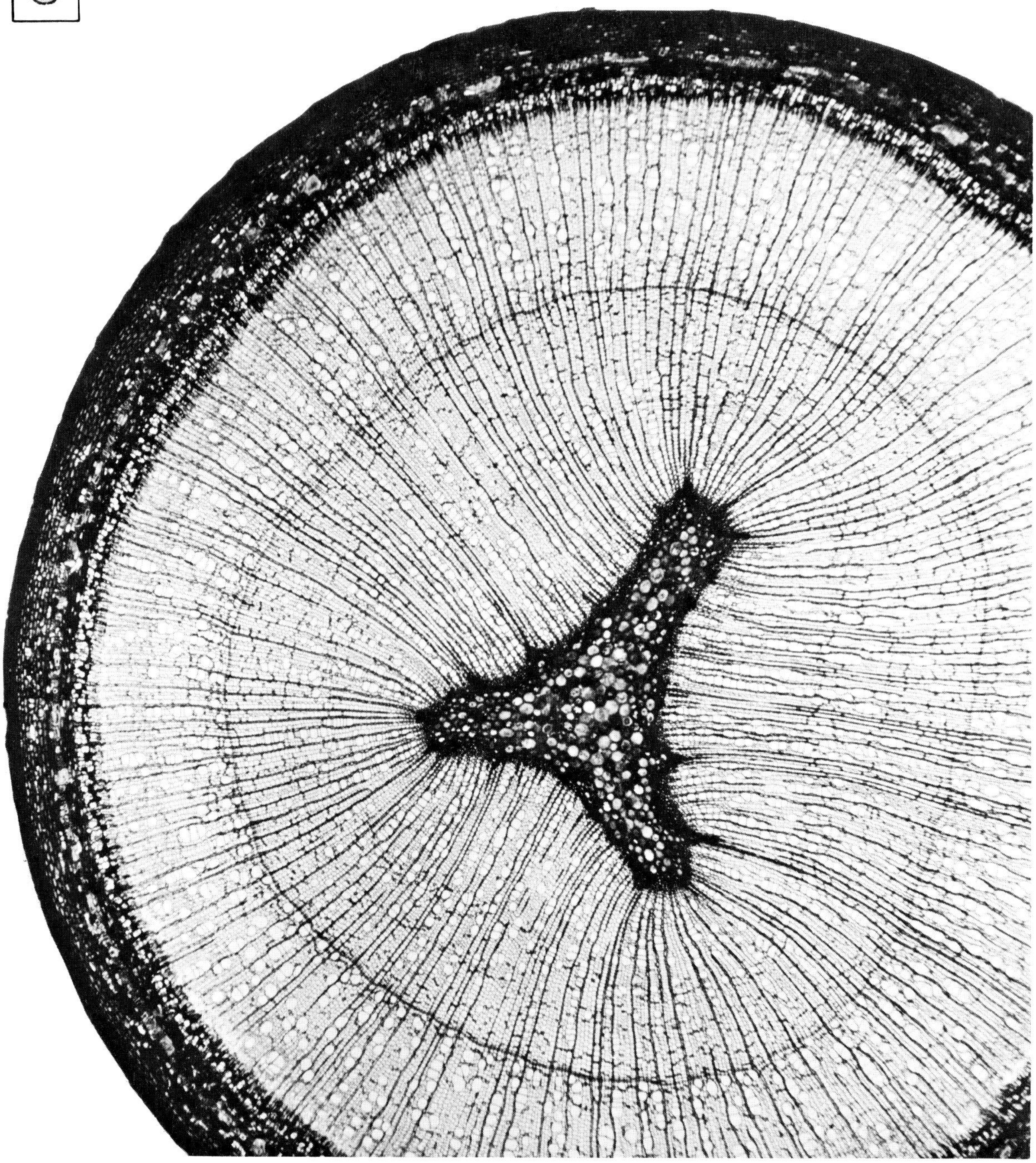

"The Staple of the stuff is so exquisitely fine, that no silkworm is able to draw anything near so fine a thread. So that one who walks about with the meanest stick, holds a piece of Nature's Handicraft which far surpasses the most elaborate Woof or Needle-Work in the World."

—Nehemiah Grew 1682—

INSIDE WOOD

MASTERPIECE OF NATURE

By WILLIAM M. HARLOW, Ph.D.

Emeritus Professor of Wood Technology
State University of New York, College of Forestry

"Human ingenuity can never devise anything more simple and more beautiful, and more to the point than nature does."

—LEONARDO DAVINCI

Wood

"A piece of wood is a wondrous thing!",
Many will often say,
And to this fact I don't object
In any single way.
They'll talk forever about its strength
And versatility—
They'll paint it, carve and cut it;
Yes—They go on endlessly.
With it they'll build many homes,
From the fibers, paper they'll make;
They'll burn it and they'll bend it—
Even art forms of wood they'll create.

Now, I appreciate technology
In every form and way,
And I'm thankful for the progress
It gives to life each day.
But I desire to go outdoors
And view an ancient glade,
And sit and stare at greenish guards
That make the forest's shade.
For in my heart I treasure wood
As it's naturally found to be:
Unheralded and priceless—
Just holding up a tree!

—MICKEY SULL

Printed
in
U.S.A.

To Charles, My Son

Preface

Man, trees, and wood have been constant companions ever since man's beginnings in the distant shadowy past, some one million years ago. The trees were here first, having evolved through a period of 150 to 300 million years.

As a forest and tree dweller, man's hands were accustomed to grasping branches in climbing, and his children ever since, if given a chance, find great sport and deep satisfaction not only in climbing trees but in building tree houses in their strong spreading branches.

During one or more periods, the climate may have become drier, and forests decreased while grasses covered more of the land. An erect creature would have the advantage of being able to see over the tops of tall grasses. In any case, gradually, man became erect and developed the typical heel and toe stride not found in any other of earth's creatures.

It is interesting that just as man is the only truly erect walker among the mammals, the tree is the only upright plant of any great height in the plant world. Consider too that both trees and men have "trunks" and "limbs"!

This book is an exploration of the beautiful structures of wood as seen through a hand lens, the light microscope, and finally the electron microscope. After the structure of wood is understood, its unique physical and chemical properties may be touched upon and illustrated.

It may seem unlikely, but it is true that the average modern man knows little more about wood than did his Stone Age ancestor of a quarter million years ago. Present day man is surrounded by wooden structures, objects and products, at least 5,000 kinds of them if paper is included, but the essential nature of wood escapes him and for a very good reason. His eyes simply cannot see what a piece of wood really is, its structure is much too fine for his powers of vision.

The unusually large number of illustrations provides the unique feature of this book. In the field of structure, to see is to understand, and wood is the most beautifully complex structure produced by plants. No two pieces are ever exactly alike and yet the patterns are similar enough so that they can be recognized with hand lens and microscope.

This book would have been hardly possible except for the help of a number of my colleagues and staff of the New York State College of Forestry. They were always ready to answer my numerous questions and to direct me to specific papers in their fields of research. Several of them also read portions of the manuscript. I have not dignified them with their titles but most are Ph.D's. and known nationally and internationally for their research on wood. To all of them I express my deeply felt thanks. M. M. Alexander, E. A. Anderson, A. H. Bishop, F. E. Carlson, W. A. Côté Jr., R. W. Davidson, A. C. Day, C. H. de Zeeuw, G. F. Earle, C. E. Farnsworth, R. C. Hartenstein, E. H. Ketchledge, N. P. Kutscha, G. H. Kyanka, R. F. La Plaine, R. V. Lea, J. A. Meyer, R. A. Moore, F. W. O'Neil, S. W. Potter Jr., Conrad Schuerch, J. F. Siau, S. B. Silverborg, C. Skaar, G. H. Smith, T. E. Timell, H. E. Wilcox.

The photo-illustrations of hemlock, southern yellow pine, redwood, sycamore, and red oak on pages 2, 4, 13, 14, and 16 first appeared in PATTERNS OF LIFE by W. M. Harlow, Harper and Row, New York City, 1966. Those of hemlock and ponderosa pine, page 6, are from TEXTBOOK OF DENDROLOGY 5th Ed. by W. M. Harlow and E. S. Harrar, McGraw-Hill, Inc., New York City, 1968. The photo-models, pages 10, 11 and 12 are by A. C. Day.

Table of Contents

◀ **Figure 1 Red pine cross section × 1½**

Chapter I

The Story of Tree Rings

The growth layers of temperate climate trees show as rings when viewed on a stump or piece of wood cut crosswise (cross section). The growth history of a tree is recorded in its rings. Some ring patterns are easy to read, others are often puzzling. In the picture, each *growth ring* consists of an inner light colored part, and an outer much darker portion. The inner part is the rapidly growing *earlywood* (springwood), and the outer is slow growing *latewood* (summerwood). In red pine and most other hard pines, the change from earlywood to latewood is quite abrupt.

Growth rings are often called annual rings, but in some trees an extra *false ring* or two may be included in one year's growth, and in other trees (see redwood p. 13) *discontinuous rings* may be formed which extend only part way around the tree. In either case one may be led astray in estimating the age of the tree. Also growth rings indicate the age of the tree only at the *height* where the cross section is made. On a stump, the years it took the little tree to reach this height must be added to get a true age count.

This red pine section is a classic example from a tree grown in a plantation with others around it, perhaps six feet apart. Growth was rapid for about 10 years. Then the crowns of the young trees came together and began to shade out the branches below. Also the network of roots needed more and more moisture, and the trees competed for this as well as for light overhead. For some years the rings got narrower as competition increased, and the whole plantation stagnated. Then a release thinning was made, and the trees quickly responded. If this thinning had been made when the crowns first began to close, good growth would have continued. This is the problem of the forester, and it is a problem because it costs money to thin, and often there is no market for the thinnings.

Wood quality is often correlated with ring width, and "too wide" growth layers in pines and other conifers often yield wood that shrinks lengthwise much more than "normal wood". Notice the very wide rings around the center of the section pictured. This "juvenile" wood is inferior to that produced from about the sixth year onward. Another thing to remember is that a narrow ring farther out toward the bark actually has a much higher volume of wood than a wide ring near the center of the tree. You can figure out how much by comparing the areas of concentric rings.

Although we know that available *light* and *soil moisture* greatly affect ring width, under Arctic conditions, especially, summer *temperatures* are often the most important factor. Available *minerals* drawn in with the water by the roots must also be sufficient in both kind and quantity. Since food making is done in the leaves, repeated *de-foliation* by insects or anything else will cause narrow growth rings. Another thing that will make for narrow rings is damage or removal of the top of the tree, especially the terminal leader. Here, auxins are formed which, carried downward, stimulate growth along the tree trunk.

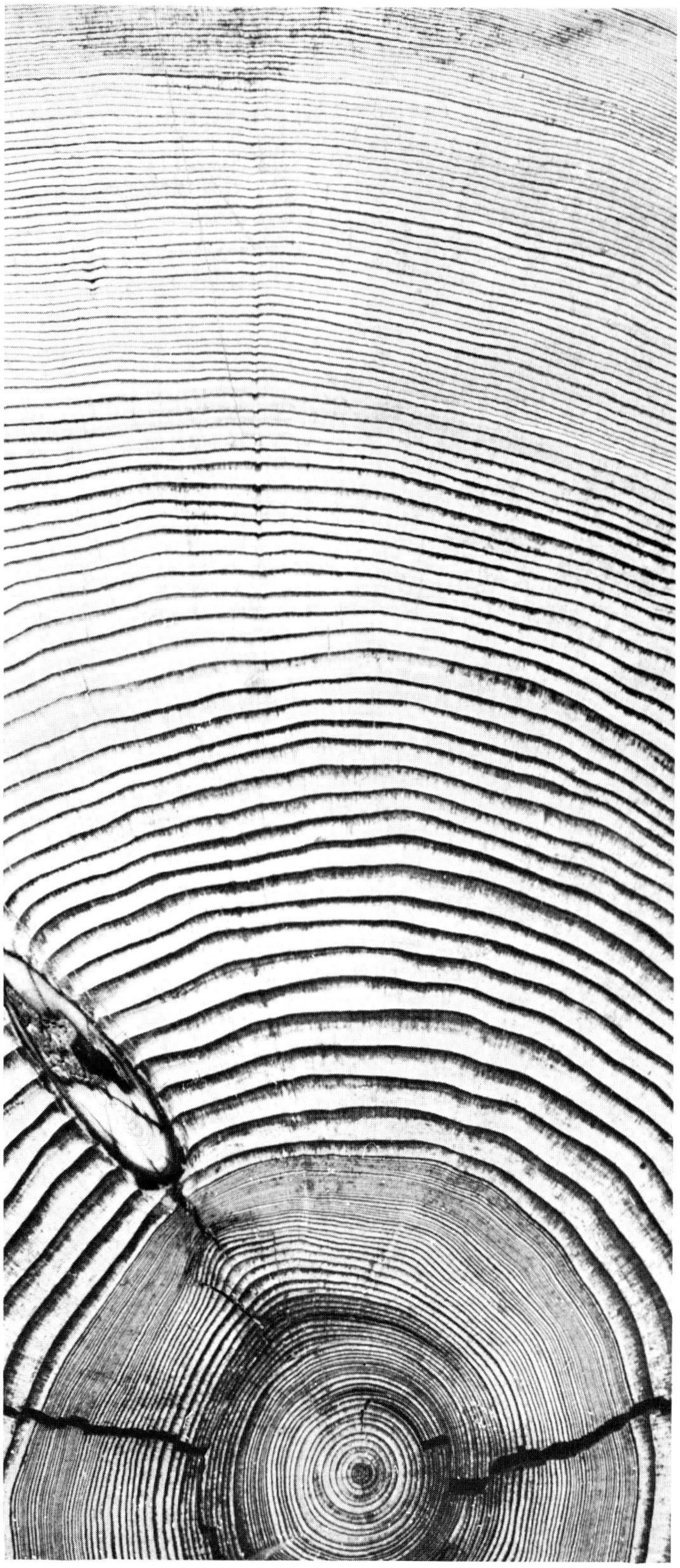

Figure 2 Eastern hemlock cross section × ¾

Figure 3 Scotch pine cross section × 2

One could not imagine a much greater difference between the growth patterns of two trees than that shown by the hemlock and pine illustrated.

Hemlock is a very tolerant tree. It will grow under dense shade for many years. This tree grew from a winged seed that, slipping from the opening cone high up in the crown, came sailing and spinning down through the twilight under a heavy forest cover. The seed landed on moist humus, and lay dormant through the winter until spring should sound her trumpet over the dreaming earth.* Then a tiny root emerged and pushed its way down through the humus. A slender stem bearing three seed leaves struggled upward. In succeding years the little tree grew steadily but very slowly in the dim light of the ancient forest not yet lumbered by man. Look at the ring pattern carefully. Near the outer edge of the central core of slow growth, the rings are so close together that you cannot count all of them. In all there are about 76, and the tree was only some four inches in diameter. Then came the lumberjacks who cut the big timber. Perhaps in a few days' time, the whole picture changed, and for the first time in its life the tree had more than enough light. Also as the roots of the felled trees died, competition for water and minerals lessened. Within a year or two, the growth increased enormously, with rings ten times as wide as those formed when the tree grew in dense shade.

Toward the top of the picture, the rings are narrower again, probably due to shading once more, and the advancing age of the tree.

The Scotch pine section is from a tree grown to planting size in a nursery, and then planted with more than a thousand others like it in an old field of an abandoned farm. From the very start the tree had full light, and its growth rate (note that the picture is twice natural size) shows that it made the most of its favorable environment, and grew steadily until a group of several narrow rings near the top indicates a depression. After this was over, normal growth resumed. What caused the narrow rings?

Back in the 1930's there was a Civilian Conservation Camp at the corner of the old field, and the campers planted the trees. Years later (about 1954) a family of porcupines took residence in the old latrine, and destructive animals that they are, they climbed some of the pines and sat there by the hour munching the bark down to the wood. Enough innerbark and cambium were removed to cripple the trees for several years.

The tree illustrated was cut in March, 1967, and the last growth ring was 1966. This and the 1965 ring have been trimmed from the top of the picture. When the porcupine damage was done, the tree was girdled about six feet from the top—all of which died. This robbed the tree of about one half of its crown.

In 1951, the crowns of the trees were closing, so a thinning was made to give the remaining trees a chance to continue their good growth. Had this history not been known, one would have to guess the reason for the group of narrow rings.

*Adapted from "Ode to the West Wind" by Percy Bysshe Shelley.

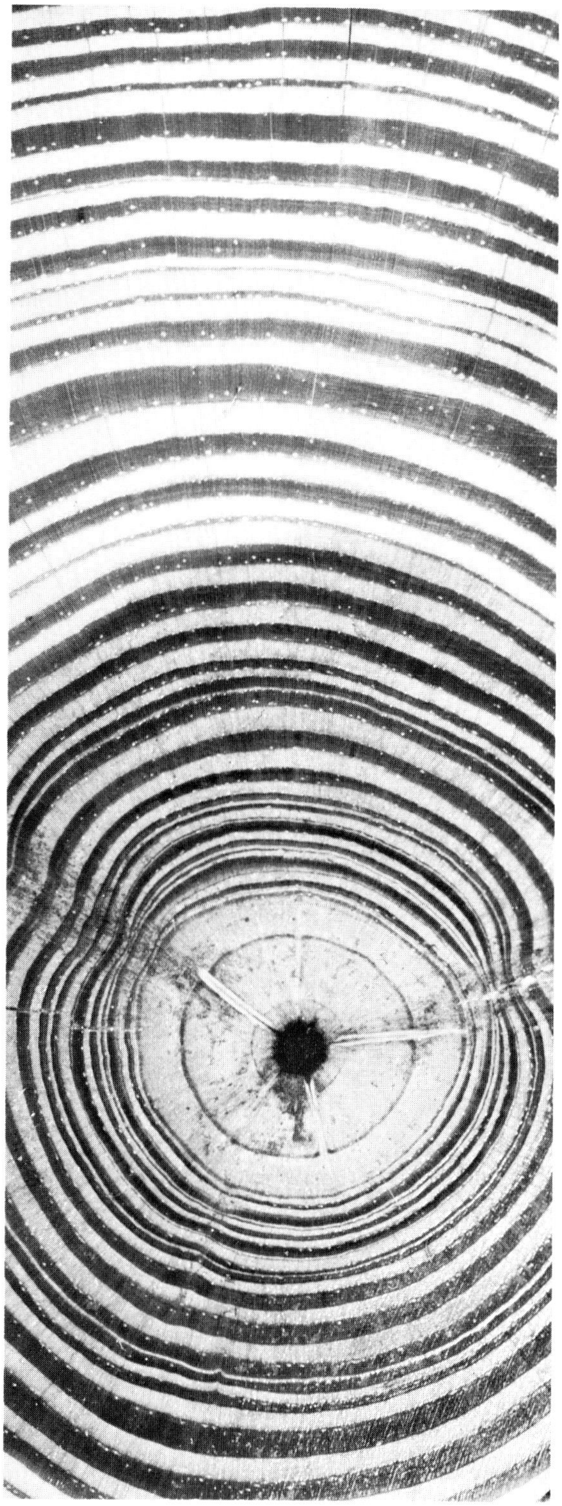

Figures 4 and 5 Southern yellow pine cross sections × 2

These two samples were picked out of the trash barrel in the New York State College of Forestry's wood working shop, but they might have been found in a lumber yard, around new construction, or in other places. Again they show that every piece of wood bears in its rings its past history.

The piece on the right grew quite evenly at the start. The relative narrowness of the rings might suggest that the tree was shaded. After the eighth ring which is variable in width, growth accelerates, but slows again near the top of the picture.

The real puzzle is the piece on the left. Notice how growth boomed during the second and third years, and then suddenly dropped sharply. Very narrow rings were formed for several years, and then the tree recovered from its depression. What could cause such a sudden drop-off in growth? Since all the wood in a tree is built up by food (mostly sugars) made in the leaves (needles in pines), perhaps we should wonder what happened to the food making process. It is highly unlikely that the tree could have been shaded so suddenly by other trees around it, and where these pines grow there is plenty of rainfall and soil moisture.

We must conclude that something, between the laying down of the third and fourth rings shown, removed most or all of the tree's needles. The two probable choices are insects and fire. It is suggested that a fire swept through the area where the young tree was growing and burned off its needles. Recovery was slow, but from the middle of the picture upward the tree again is growing normally.

These few examples of ring patterns are enough to show how sensitive a tree is to all the conditions surrounding it. Best growth is made when there is a balance of many factors some of which may not yet be known.

Each ring pattern is unique, no two are ever exactly alike. They are as individualistic as finger prints. You will find that every piece of wood you pick up has the story of its life through the years, forever imprinted upon it.

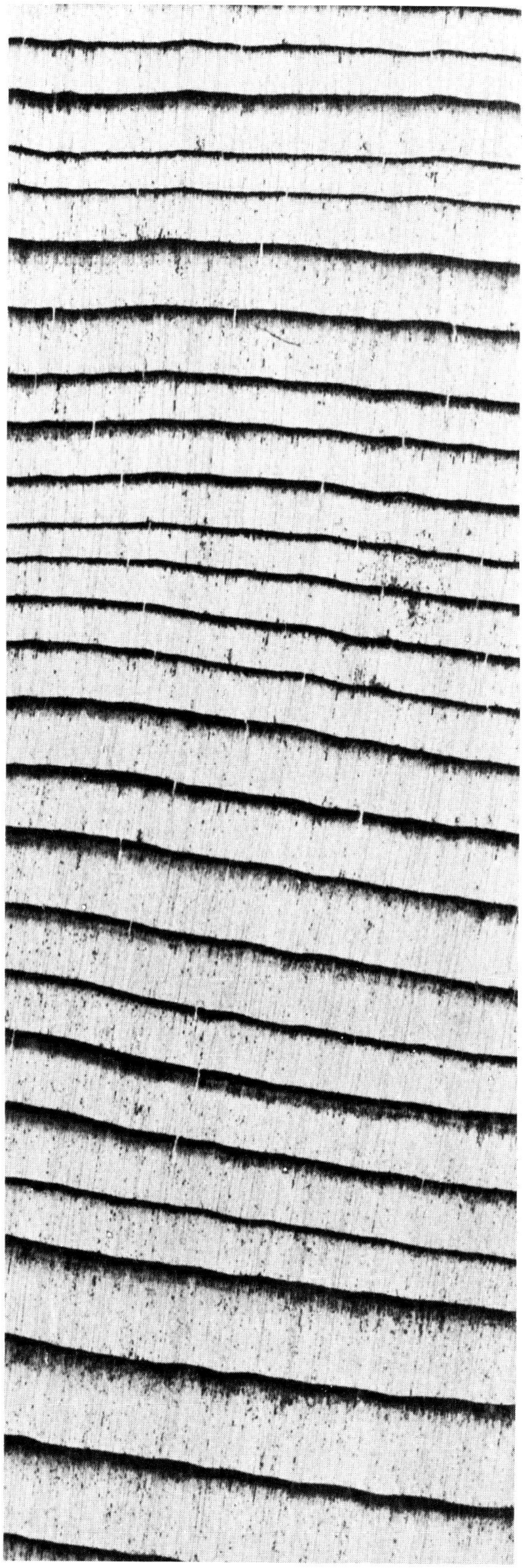

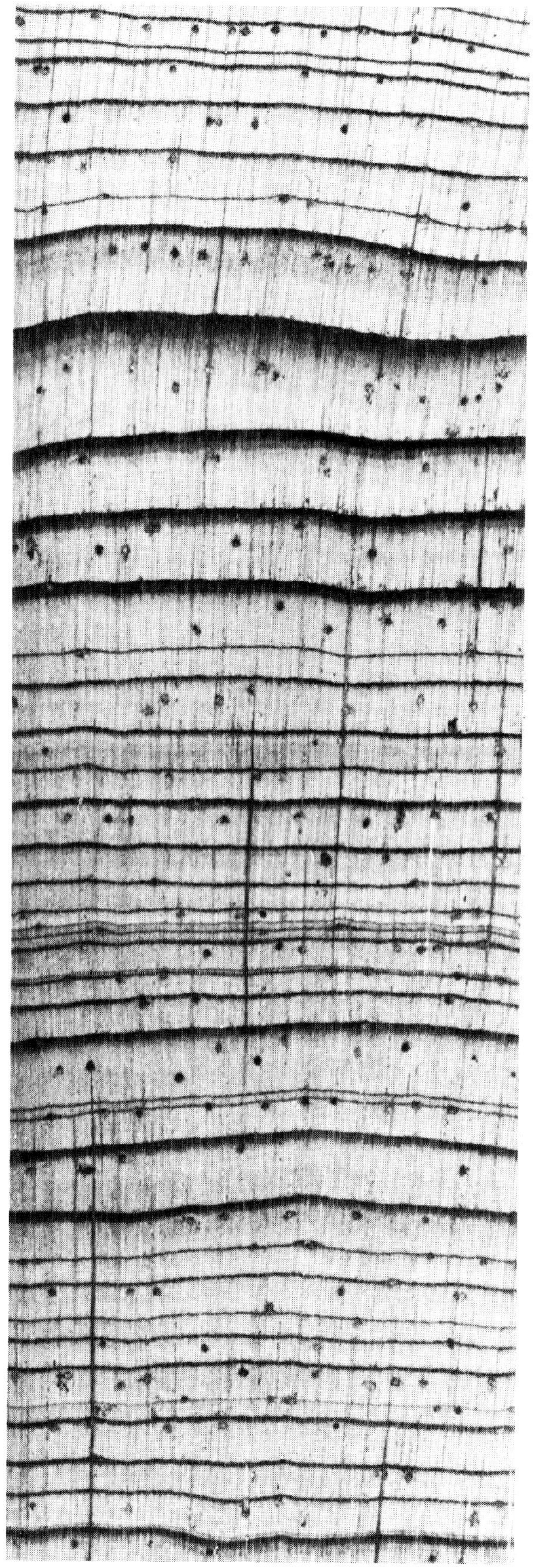

Figure 6 Eastern Hemlock
Figure 7 Ponderosa Pine
Wood Cross Sections × 6

Dendrochronology

"Dendro" is a word element meaning "tree", "chronology" signifies time, and dendrochronology is the science of tree rings and what they tell of past history. Ring width depends upon several factors (p. 1). If we could assume that all but one of these factors remained more or less constant during a tree's life, then ring width would correlate with that factor which varied significantly from year to year and at times limited the tree's growth.

In the arid Southwest, soil moisture is the principal variable. Some trees get very little of it; in critical years, often barely enough to survive. Since they are widely spaced from each other, there is little effect of shading. Such trees, especially those growing on dry slopes often show "sensitive" ring patterns which can be correlated with annual precipitation. The ponderosa pine section is from one of these trees. By contrast, the eastern hemlock ring series with its uniform pattern is called "complacent" and is useless for such studies. Year after year it had enough water, light and other things to grow steadily.

Look again at the ponderosa pine section. There is great variation in ring width; the widest ring is about 18 times the width of the narrowest ring. This does not mean that during the year of the wide ring there was exactly 18 times more precipitation, only that there was a "great deal" more soil moisture than during the drought years.

In the early 1900's, A. E. Douglass, an astronomer, was working with sunspot cycles and their effect on climate. He wondered whether tree rings might record precipitation, and began to study them. He was fascinated by the kind of patterns shown by the ponderosa pine section, especially when he found that trees scattered over a wide area all had similar patterns, near enough alike so that they could be matched with each other. Why could not a calendar of tree rings be developed reaching into the distant past? To this idea he devoted his life, and the Laboratory of Tree-Ring Research at the University of Arizona, Tucson, is a memorial to him and his students.

To begin the calendar, he cut a living tree. If this were done at the end of the growing season, it would be easy to see the last ring formed just under the bark. He could then count and date the years in to the tree's center. Perhaps nearby there might be an old stump of a tree logged many years before. The ring pattern at the *outside* of this old tree might match that near the *center* of the living tree just cut. If so, the calendar could be extended toward the center of the older tree. By searching for still older trees, there has been developed through the years a tree ring calendar extending backward in time for thousands of years. Such tree ring calendars are used by archeologists in dating prehistoric ruins of pueblos in the Southwest, and meteorologists find them valuable in studying the climate of past ages.

On a trip to the Grand Canyon, I sawed off at random a piece from a split log of ponderosa pine in a pile of firewood. It is the one illustrated, and the Tree-Ring Laboratory found that it fitted their calendar with "textbook" perfection. The widest ring is dated 1767.

Figure 8 An embedded branch (knot) in Douglas-fir

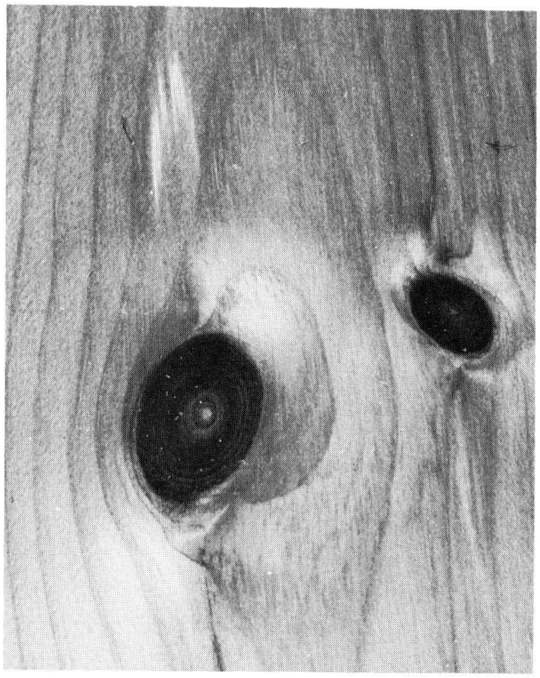

Figure 9 Cross section of two embedded branches (tight knots)

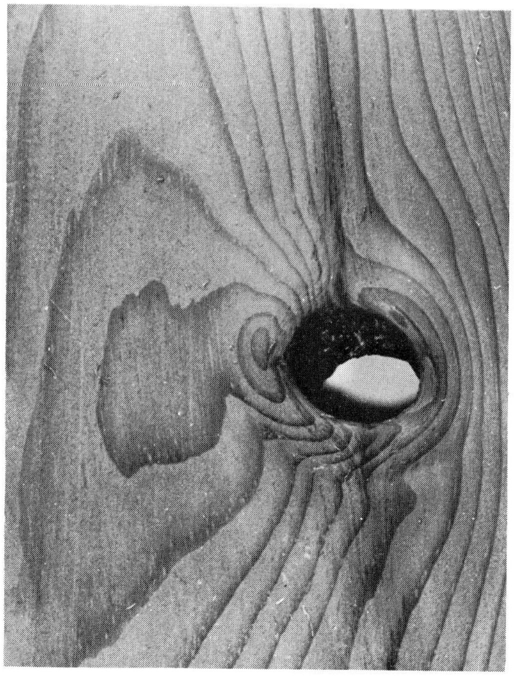

Figure 10 Cross section of a knothole.

Knots

If you ask a large number of people what causes knots in wood, you may get some strange answers. Actually a knot is simply an embedded branch in the wood of the tree trunk. Later, we shall explore in detail the structure of the *cambium,* the remarkable layer of live cells between the wood and the bark. Each growing season, the cambium, by cell division, adds a new layer upon the wood already formed, and also a layer of inner bark (phloem) on the cambium's outer face. The cambium extends like a glove over the entire surface of the tree's body, except at the very twig tips, and root tips. Therefore, the woody growth layers produced by it do the same. Look carefully at the upper illustration and see how the vertical growth layers of the tree's trunk curve and become much narrower to form the branch. To visualize this in three dimensions will take some imagination. This branch was pruned nine years before the tree was cut. You can, by studying the over-riding growth layers, see about how many years it took to cover the branch stub with new wood. After that, of course, the wood formed was "knotless."

The cross sections of knots, and a knothole need further explanation. As we have seen above, the branch is a living part of the tree and growth layers are added to it in the same way that they are along the trunk. This, however, lasts only as long as the branch lives. In open grown trees, the side branches live for many years, even sometimes for the life of the tree. By contrast, trees growing close together produce smaller branches and within a few years they begin to die off at the trunk because of shading. The cambium of the tree trunk keeps forming wood out along the base of the dead branch but there is no longer any living connection between the two. While the branch was still alive it was part of the tree and boards cut including it would show "tight" knots like the two in Fig. 9. As soon as a branch dies, its connection with living tissues being lost, it becomes a "loose" knot. It will fall out, or is easily pushed out of the board containing it, leaving a knothole (Fig. 10). For general purposes, a few tight knots here and there do not reduce the value of lumber a great deal, but loose knots or knotholes do.

For many years, pruning has been done, especially in close-spaced plantations of conifers. Only selected fast growing individuals are worth pruning, and usually only the first 16 feet of the trunk (the first log) is cleared of its side branches. The resulting clear (knotless) wood brings a higher price than the "knotty stuff". However, in recent years, softwood lumber with tight knots has found a special use for interior siding, and other decorative purposes; knots give "character" to wood. Knotty pine is no longer considered fit only for box boards.

Chapter II

The Structure and Microstructure of Wood

Three Dimensional Models

These block models to show wood structure in three planes of section are made by preparing photomicrographs (with a microscope) from thin sections cut from the same block of wood. The cross section is made first, but it is practically impossible to get the other two sections to match perfectly with the cross section and themselves. The small wood block must be turned in the jaws of a vise in the cutting machine for each of the vertical sections, and alignment of the surface with the cutting knife results in a wasted section or two before a good one can be made. However, you will have no difficulty matching up the various structures shown.

HEMLOCK is a conifer or "softwood," and like others in this very ancient group of trees, the wood structure is relatively simple. Mostly, the cells (*tracheids*) are about 3.4 mm. long making them some 100 times longer than their diameters as seen in cross section. Look carefully

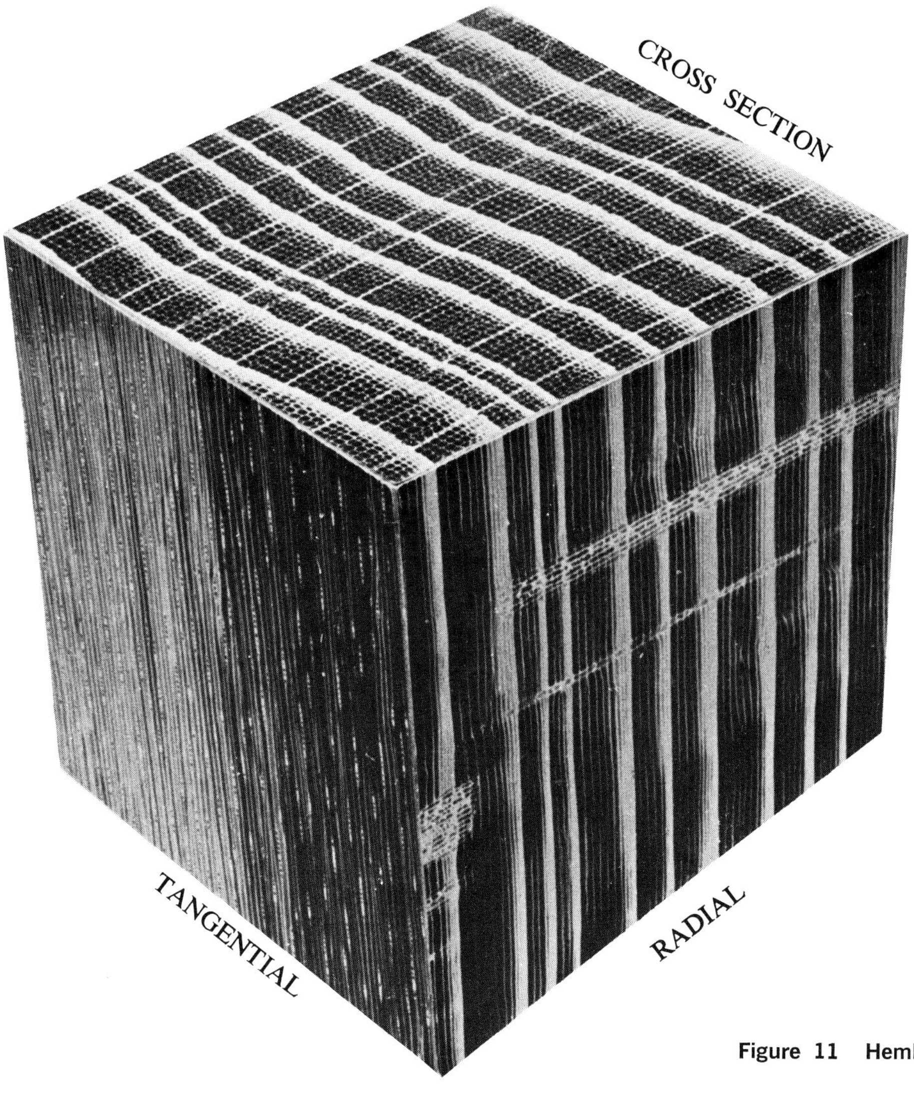

Figure 11 Hemlock photo-model × 30

Figure 12 Birch photo-model × 30

at the edge where the cross section meets the radial section, and then down along the radial face. At this magnification these cell details are just about visible. Later in the book they will be shown much larger, but here you can still see the growth rings, and begin the transition to wood structure with the microscope. Notice that the cell walls in the earlywood tracheids are so thin that they barely show. In the latewood, the walls are very thick and the cell cavity (lumen) is very small. Which type of cell is best for conducting sap up the tree? Which one gives the most strength to the wood?

One other structure shows, the *wood ray*. In cross section, wood rays appear as fine lines crossing the growth rings at right angles. On the radial face, they are seen as ribbons of varying width made up of short brick-shaped cells running horizontally.

Pines, spruces and a few other conifers show also large-to-small holes in cross section, and in freshly cut sapwood, resin oozes from them. These *resin canals* are scattered in the cross section, and run lengthwise in the tree.

BIRCH represents the broad-leaved trees, or "hardwoods." You will see at once the large *pores* in cross section which open into long tube-like *vessels* that run lengthwise in the wood. Most of the conduction of the sap upward from the roots is through these large vessels, and the *fiber-tracheids* serve principally as strengthening tissue. Because of this specialized conduction system, the hardwoods are considered more "advanced" than the softwoods. In birch and

Figure 13 Ash photo-model × 30

many other woods, the pores are scattered throughout the growth layers. Such woods are *diffuse porous.*

ASH is also a broad-leaved tree but the wood is quite different from that of birch. In ash the first vessels formed in spring are enormous compared to those in the latewood which is mostly made up of heavy-walled fibers. Such *ring porous* woods show a further development in the separation of conductive and strength-giving functions into separate tissues. Wood rays are features of both softwoods and hardwoods.

Wood structure will be shown in much greater detail later, but this much will give you an appreciation and understanding of the section which follows.

Figure 14 Redwood × 5. This cross section must have come from a very large (note slight curvature of growth rings) and very old tree, perhaps 1,000 years of age. Careful observation will show great variations in growth over the some 130 years recorded. The struggle to grow is dramatically shown. ▶

Some Woods (Cross Sections) at Low Magnification

Figure 15 Sycamore. Cross section × 5. The small pores scattered through the growth rings (diffuse porous) are barely visible at this magnification. The wood rays are wide and closely spaced.

Figure 16 Honeylocust. Cross section × 5. The large sap conducting vessels of the early-wood contrast with the dense tissues of the latewood in each growth ring.

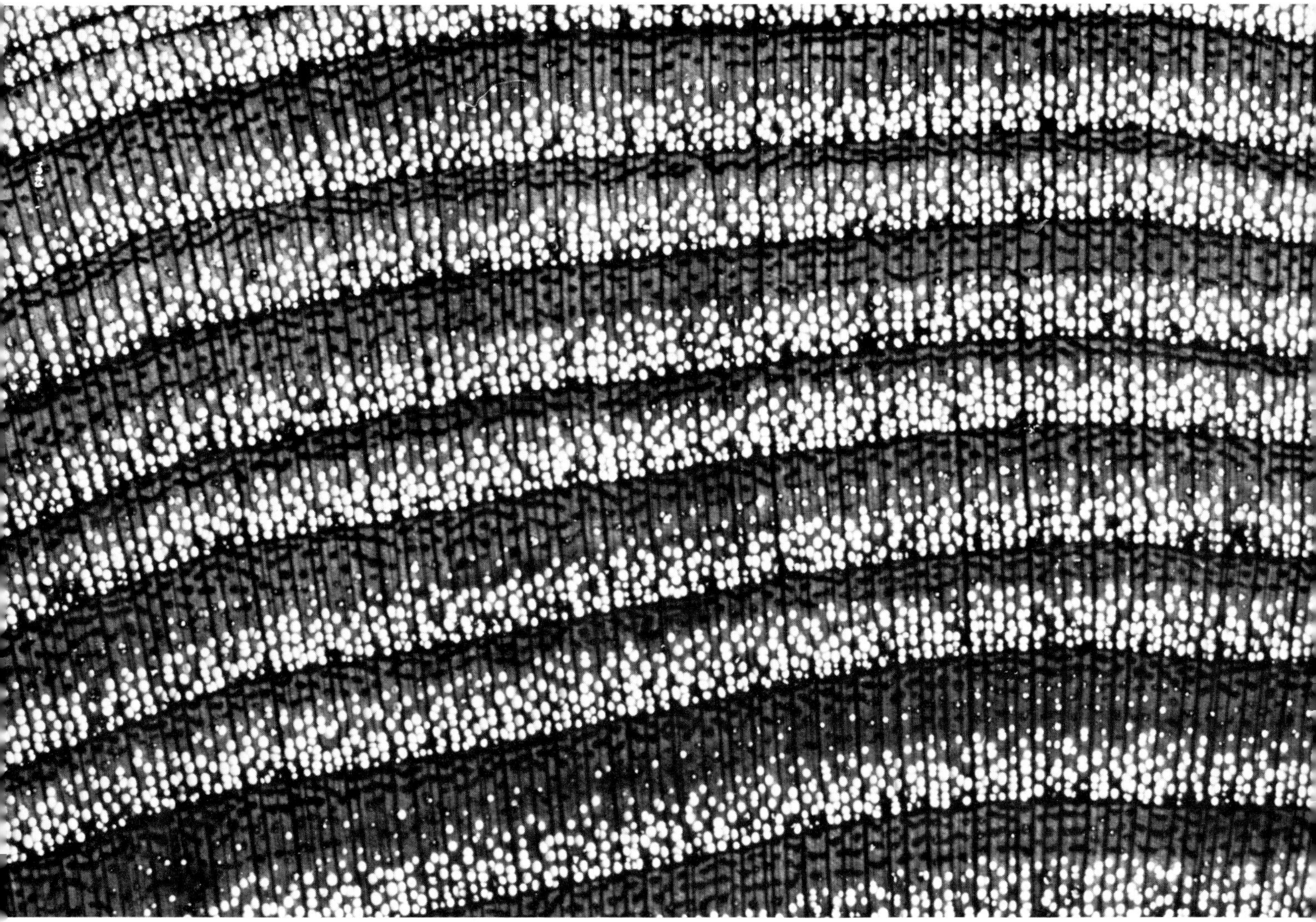

Figure 17 Red oak × 5. In this cross section the growth rings are crossed at right angles by the large wood rays radiating out from the center.

Figure 18 Black Cherry. Cross section of tree showing heartwood, sapwood and bark.

The preceding photographs of redwood, sycamore, and the others which follow in this part of the book were photographed at about 5 times natural size from large wood sections (2″ x 4½″) made by Romeyn B. Hough of Lowville, New York, in the late 1880's and early 1900's. Such large sections required a special machine which he designed and had built. For each species, there was a transverse, radial, and tangential section all mounted in a split cardboard, page-sized holder, the sections framed so that they could be held toward a light for study. In all, 14 magnificent volumes were issued. More recently, beginning in 1957, a second edition authored by Dr. E. S. Harrar, Duke University, has been issued by Robert Speller and Sons, New York City.

Heartwood and Sapwood

In a twig or young tree, the woody conducting cells (tracheids, vessels, etc.) carry sap from the roots upward to the leaves. As the years pass and the tree gets larger, there develops progressively outward from the pith a core in which all life ceases. Live cells in the wood rays or elsewhere, die. This core is the *heartwood* which is no longer part of the living organism, the tree. The wood outside of the heartwood is the *sapwood*. The amount of water conduction in it varies greatly in different kinds of trees and is, of course, seasonal. As sapwood becomes heartwood, darkening commonly occurs, but in some kinds of trees there is little change in color.

The dark color of heartwood is caused by one or more chemicals of many different kinds. These include tannins, dyes, and oils. Since most of these substances can be removed from heartwood sawdust by cold or hot water, or organic solvents, they are called extractives. Certain of these extractives in the wood resist decay-producing fungi, and so a fallen dead tree lying on the ground may have sound heartwood while the sapwood is completely decayed (p. 101).

Although the heartwood of a redwood or cypress may resist decay for a thousand years or more, that of many other trees including poplar, birch, maple, and basswood is little if any more resistant than the sapwood and decays quickly.

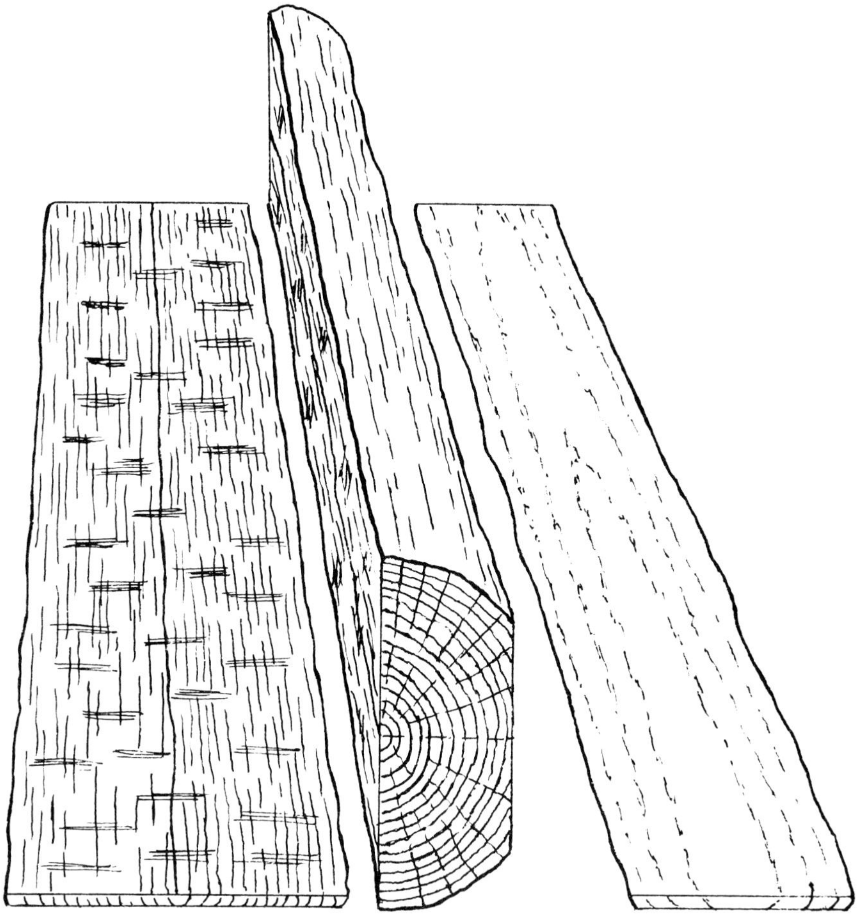

Figure 19 1. Quarter-sawed board 2. Plain-sawed board

(Adapted from an illustration in the "Wood Handbook," U. S. Forest Products Laboratory.)

Quarter-sawed and Plain-sawed Lumber

When a log is sawed through its center as shown in the illustration (1), the first board or two on each side of the center line will show the growth layers as parallel vertical lines, and the wood rays as irregular, usually interrupted, lines running horizontally. The straightness of the wood rays shown is almost never seen because it would be practically impossible to adjust a log on the saw carriage so that a cut could be made exactly through the log's central plane. Rather, the rays appear as wavy lines on the radial surface. The term *quarter-sawed* is used for such boards in the hardwoods, while a more common name in softwoods is *edge-grained*. In certain hardwoods such as the oaks with their immense wood rays, quarter sawing is done to produce the most lumber possible that shows the beautiful ray pattern on the surface. The log is first sawed into quarters, and then each quarter is positioned and re-sawed with the boards coming off as nearly as possible parallel to the rays.

Number 2 shows something quite different. Here, the boards are sawed more or less tangent to the growth rings, and an entirely different pattern results. Such boards are called *plain-sawed,* and in softwoods also *flat-sawed.* Notice that the growth layers form narrow somewhat cone-shaped patterns. Lengthwise in the tree, the trunk is a series of telescoping growth increments. This is covered in greater detail under tree growth (p. 55). If you keep on sawing boards from the outside to the center, it is clear that in the series there will be a continuing change of a surface pattern from plain-sawed to quarter-sawed. Each board will have its own unique appearance which will never be duplicated in any other board.

Figure 20 Figured Japanese ash veneer

Wood Figure

The surface pattern of a board or sheet of veneer is called its *figure*. Patterns are extremely variable, and are caused by unusual grain, texture, and color variations in a tree. When patterns are common and not especially noticeable, the wood may not be considered "figured". At the other extreme, unusual patterns of great beauty in form and color occur.

Grain in Wood

The terms "grain" and "texture" are almost hopelessly confused as applied to wood. *Grain* should refer to the direction of the fibers and other cells which make up a piece of wood. Then, such terms as straight-grained, cross-grained, and curly-grained are understandable. A special case is "silver grain" in the oaks (p. 25). For a thorough discussion of grain and texture see "Textbook of Wood Technology" (8).

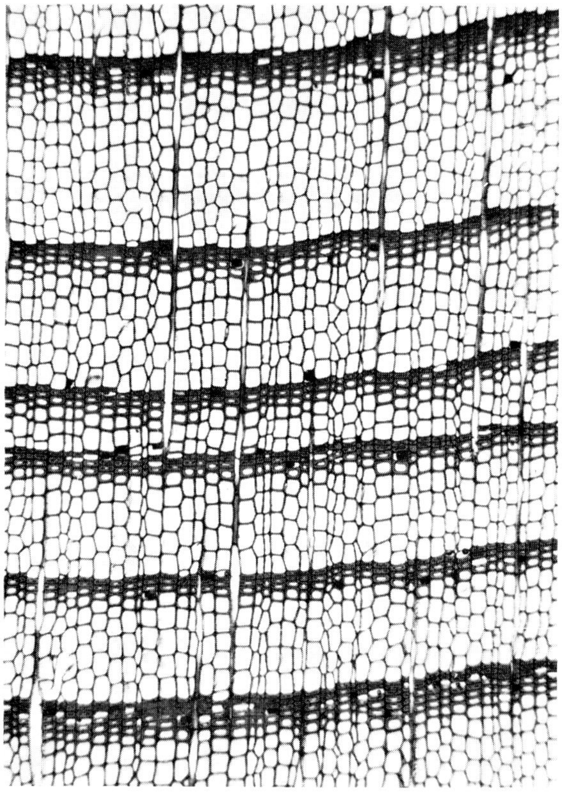

Figure 21 Redwood × 25. Sp. gr. .38

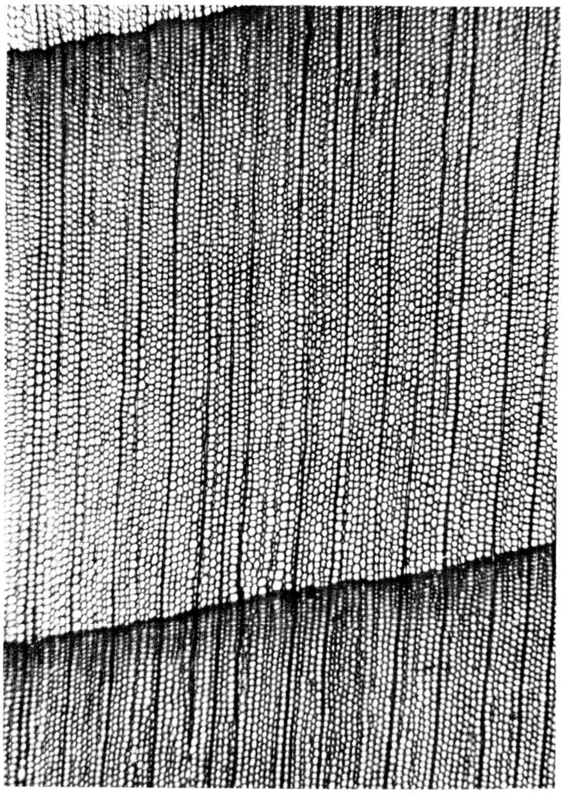

Figure 22 Yew × 25. Sp. gr. .60

Texture in Wood

The word texture is associated with textiles or woven materials, and also with the surface appearance of a substance. One could say of a fabric that it was either coarse or fine textured depending upon the size of the threads and the distances between them. The magnified cross section of a piece of wood presents an incomparable "fabric" to which the word texture may be applied. A glance shows that the redwood is coarse textured, the yew fine textured.

In coniferous woods, the *tangential diameter* of the tracheids (except near their tapering ends) is used as a measure of texture. In the hardwoods, vessel diameters, numbers and arrangement, as well as wood rays are considered. The difficulty in framing a good definition of wood texture is of course due to the extreme structural variation found throughout the thousands of tree species that have spread themselves over the earth. For more information see (8).

In any case it is interesting to see how texture influences the properties of wood. Redwood is soft, and light in weight. Yew is hard and heavy. The amazing resistance to decay of redwood is caused by tannins infiltrated through the heartwood, and has nothing to do with its structure.

Figure 23 Butternut. Tangential section showing growth layers telescoping each other

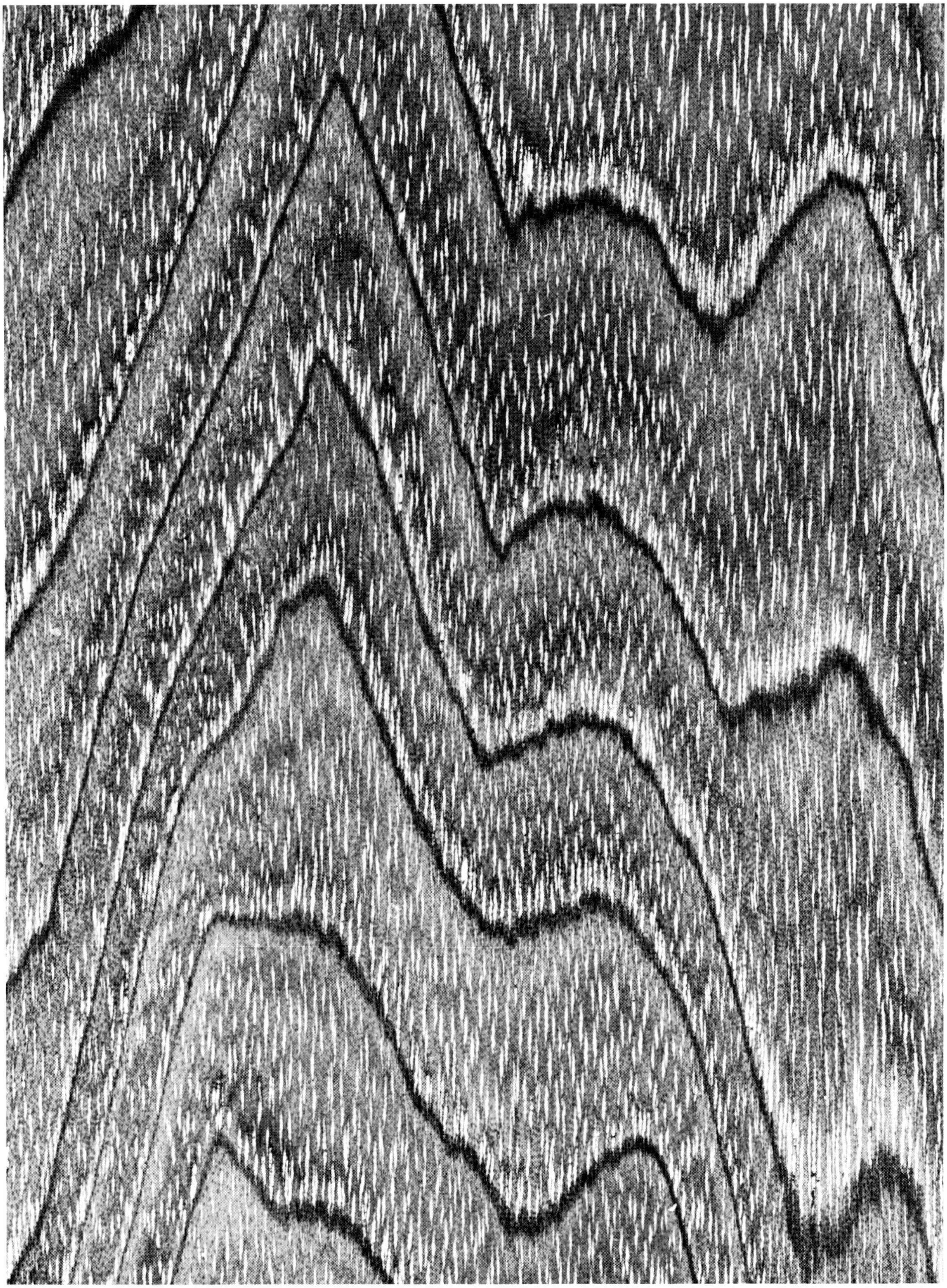

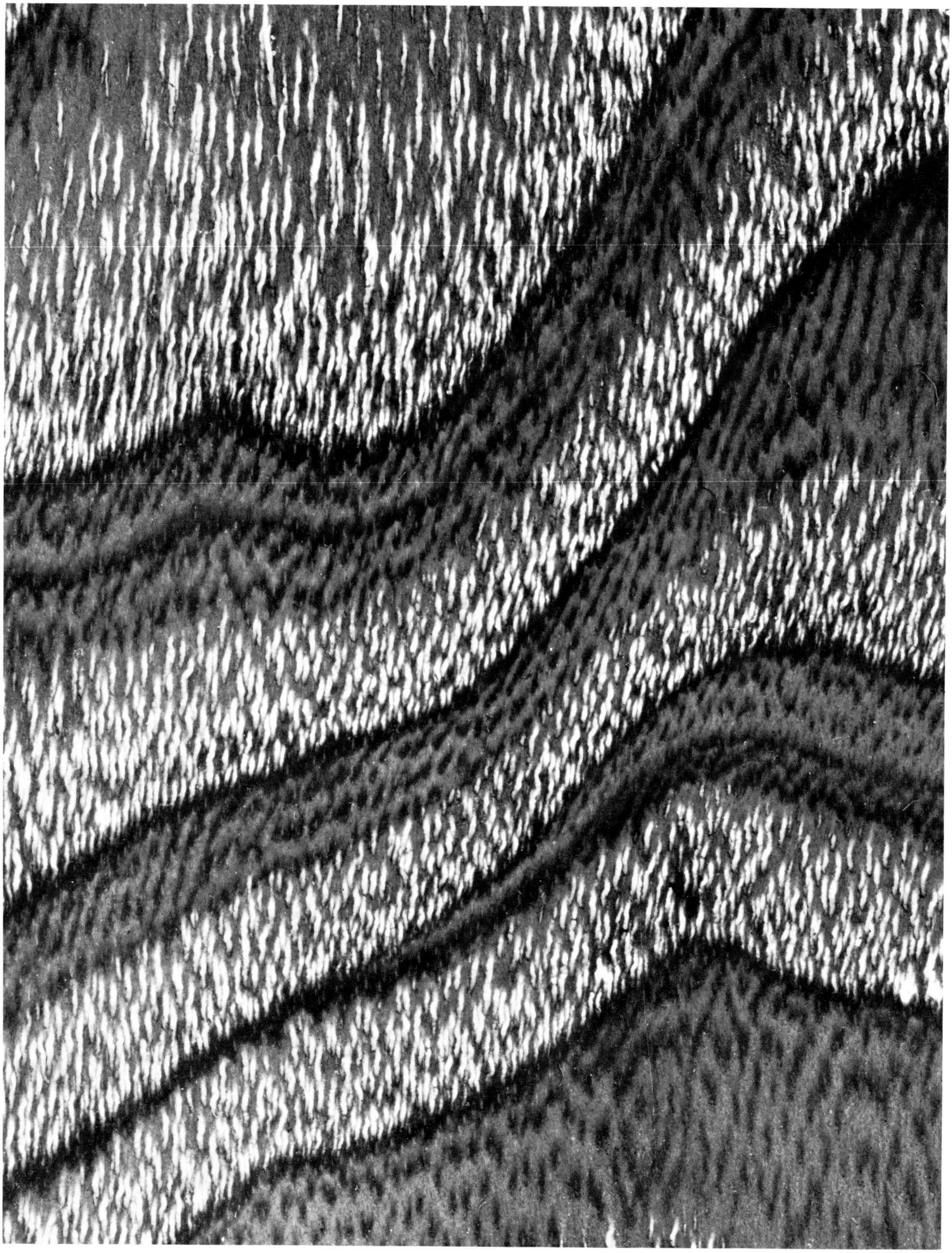

Figure 24 Honeylocust. Tangential section × 5. The earlywood vessels look like little flames

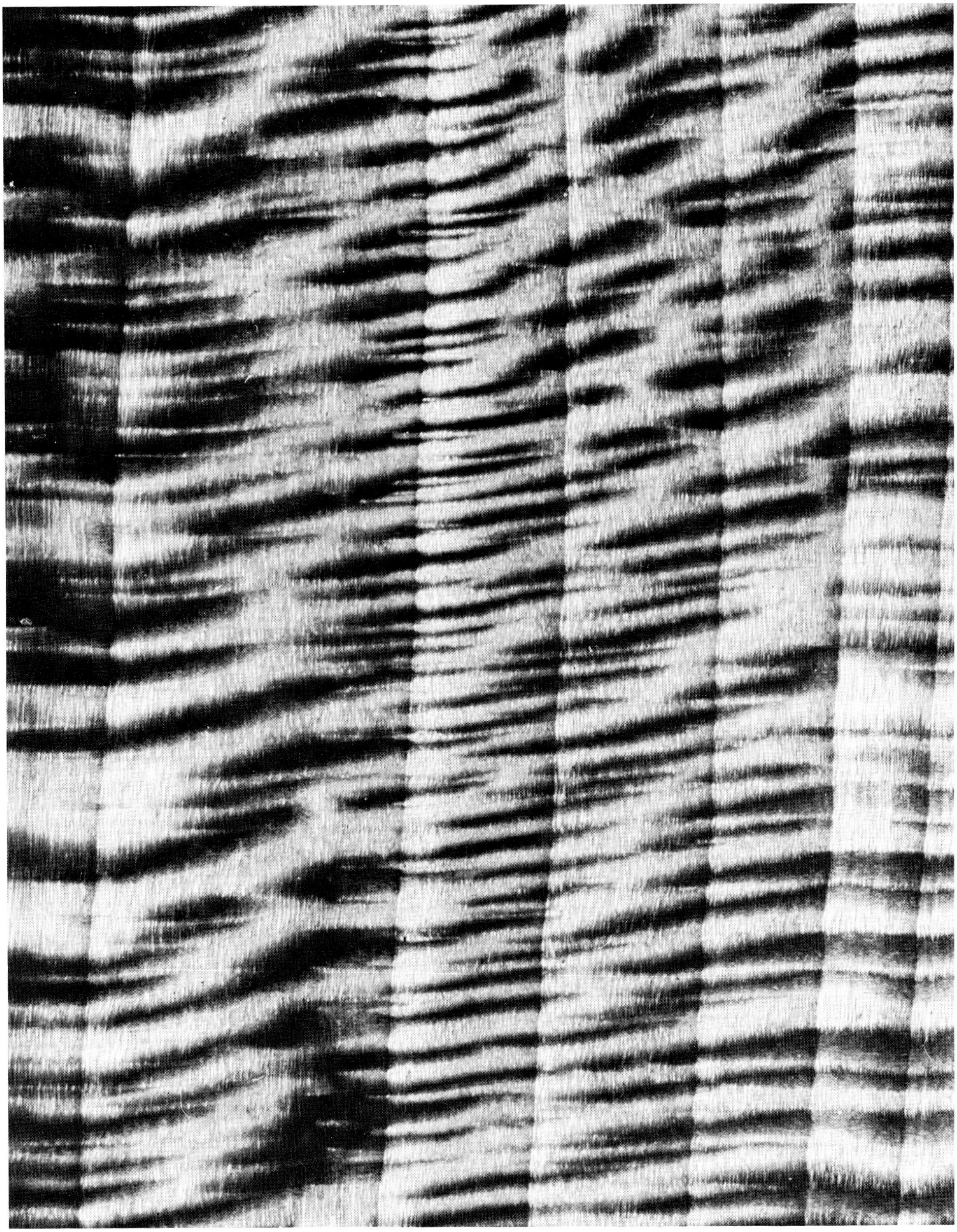

Figure 25 Sycamore. Radial section × 5. If this section were truly radial, the wood rays would appear as continuous ribbons sweeping across the page, and the effect would be less interesting.

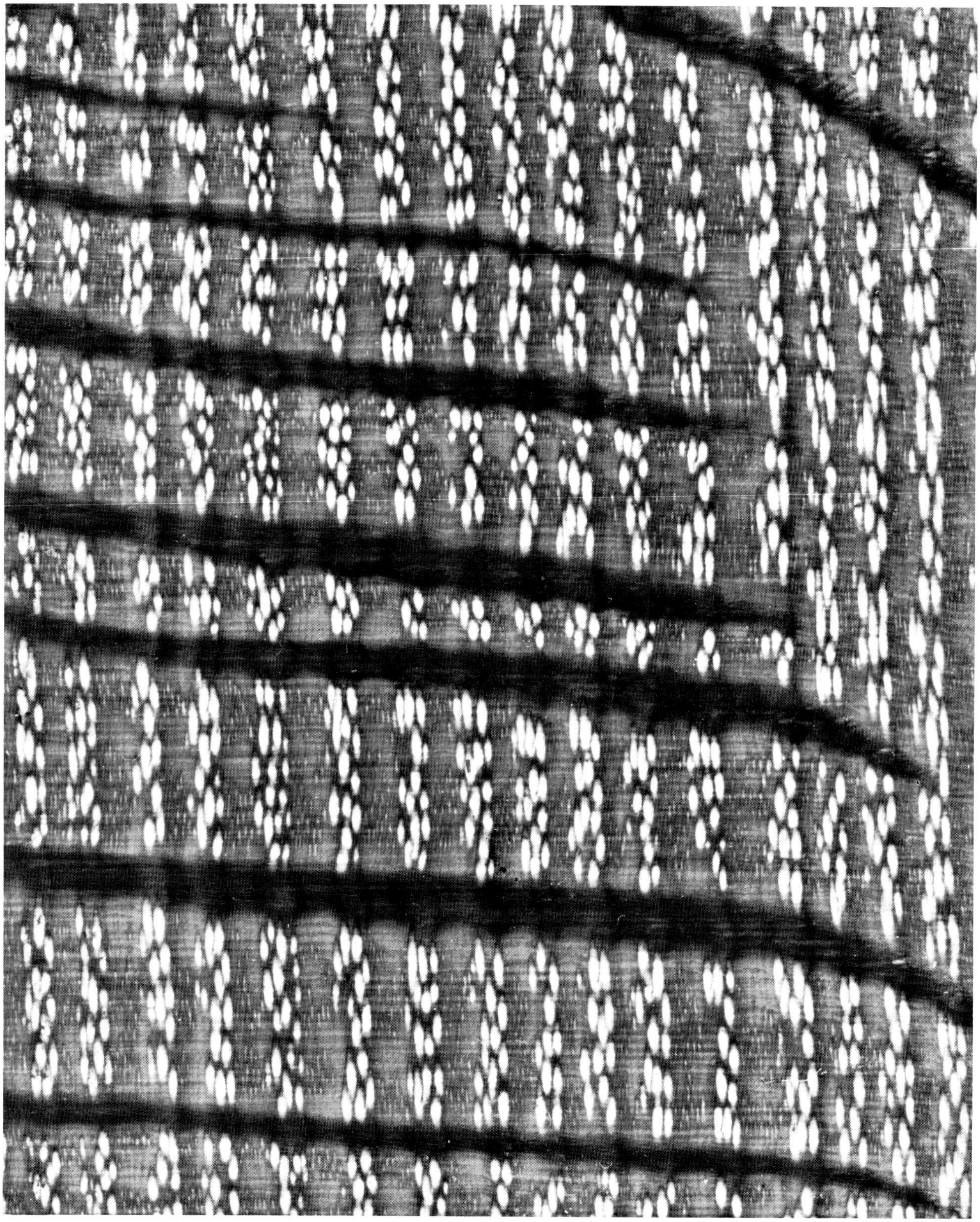

Figure 26 Red oak. Radial section × 5. Growth of this piece was much slower than that cross-sectioned (p. 16). Seventeen growth layers of alternating earlywood and latewood are shown. The vessels look short because the sectioning was at an angle and not exactly vertical.

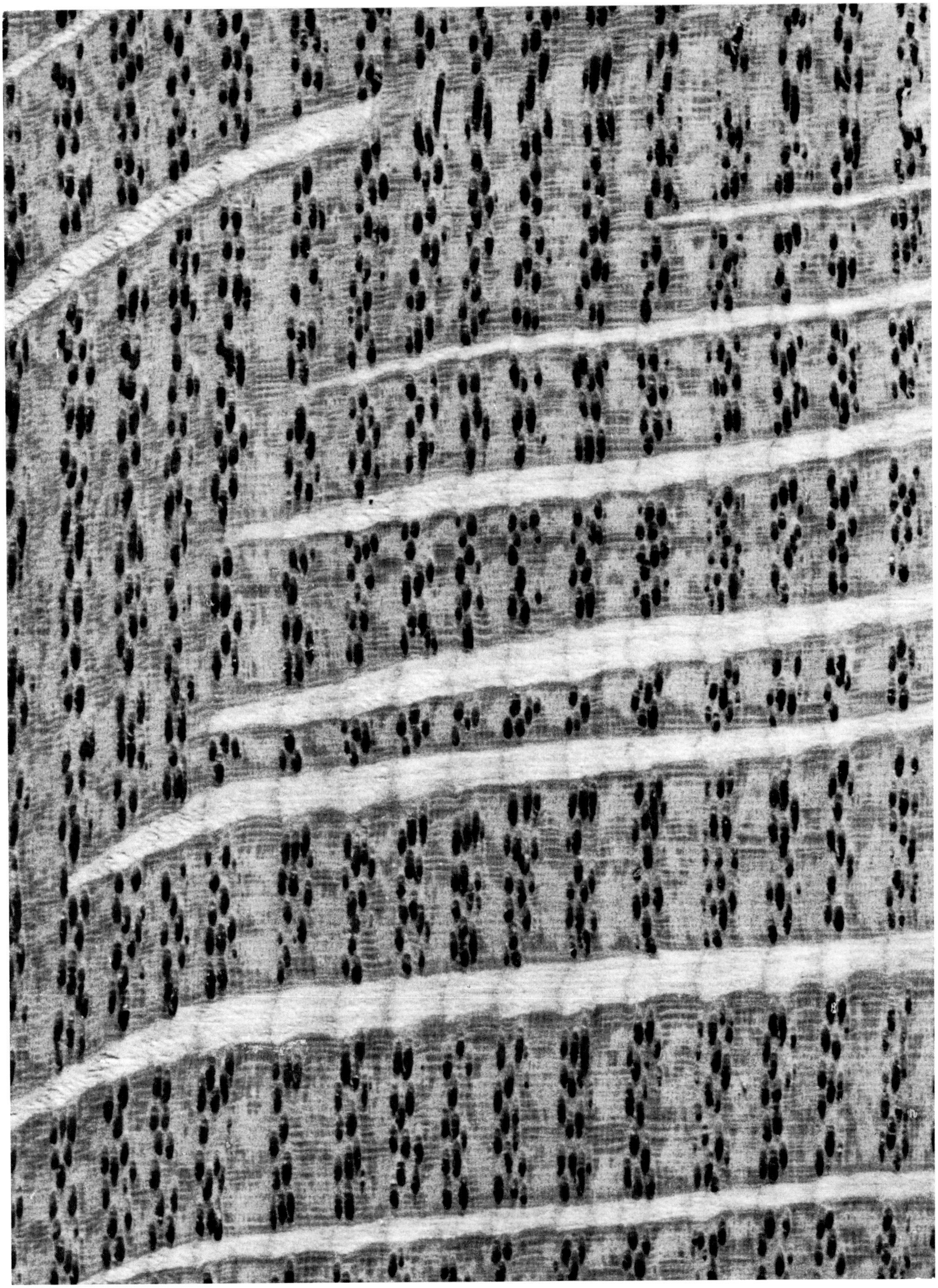

Figure 27 Same section as on facing page, but front lighted showing wood rays as "silver grain"

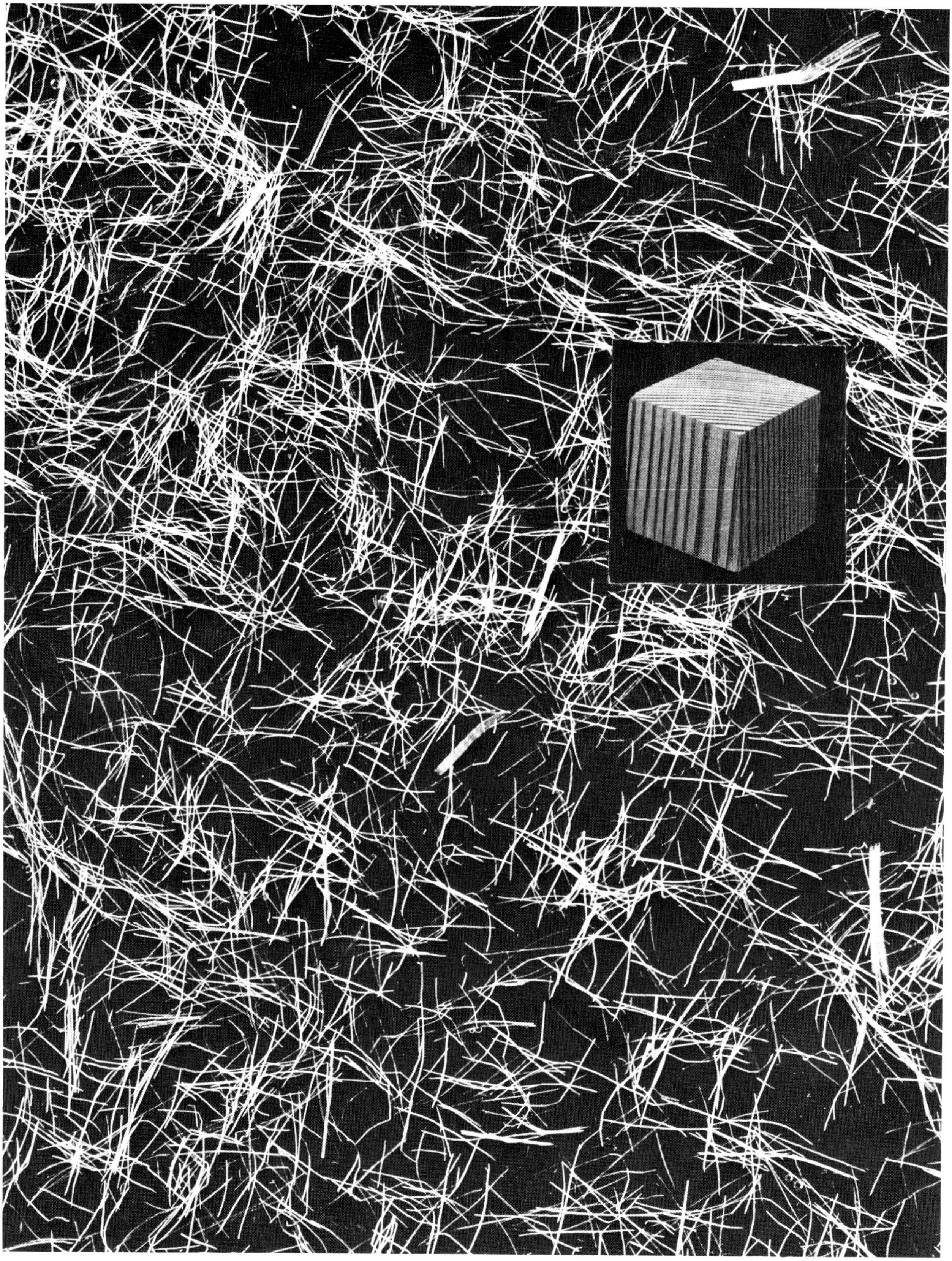

Figure 28 A one inch cube of Douglas-fir. The fibers (tracheids) of which it is made are enlarged 7 times

Figure 29 Douglas-fir tracheids × 100

Figure 30 Cut-away model, segment of a tracheid × 1600

Fiber Structure

When you hold in your hand a seemingly simple block of coniferous wood, you are actually looking at a beautifully complex structure already evolved some 300 million years ago. In Douglas-fir, except for the wood ray cells, and those forming the scattered tube-like resin canals, the wood consists of fibers (*tracheids*) in countless numbers. There are about 3 million of them in a cubic inch of the wood! Later (p. 117) we shall see how you can use chemicals to take wood apart yielding these fibers that are the invaluable raw material of the pulp and paper industry.

Appearing simple in structure at low magnifications, the cutaway model based on electron photomicrographs shows a structure of great complexity and strength. The whole cell wall is built of tiny microfibrils in layers. In the outer thin *primary wall,* the microfibrils have a random arrangement. The three inner layers together are the *secondary wall.* In the thin S1 layer, the microfibrils run nearly horizontal or at a low angle. Those of the thicker S2 layer are laid down at a steep angle, while those in the thin S3 layer approach the horizontal. Inside of the S3 layer is the cell cavity through which sap rises in the tree.

When plants grew submerged in the ocean more than 300 million years ago, they were surrounded at all times by water and minerals needed for life and growth. By degrees certain of them began to grow nearer the shore and adapted to partial drying out when the ocean retreated at low tide. On land there was plenty of room for plants, provided they could protect themselves from direct sunlight, and develop roots for getting water and minerals from the soil. After eons of time, the crowded land plants began their battle for light. This meant "going up", and involved the development of tracheids and other water conducting cells from which tree trunks are built.

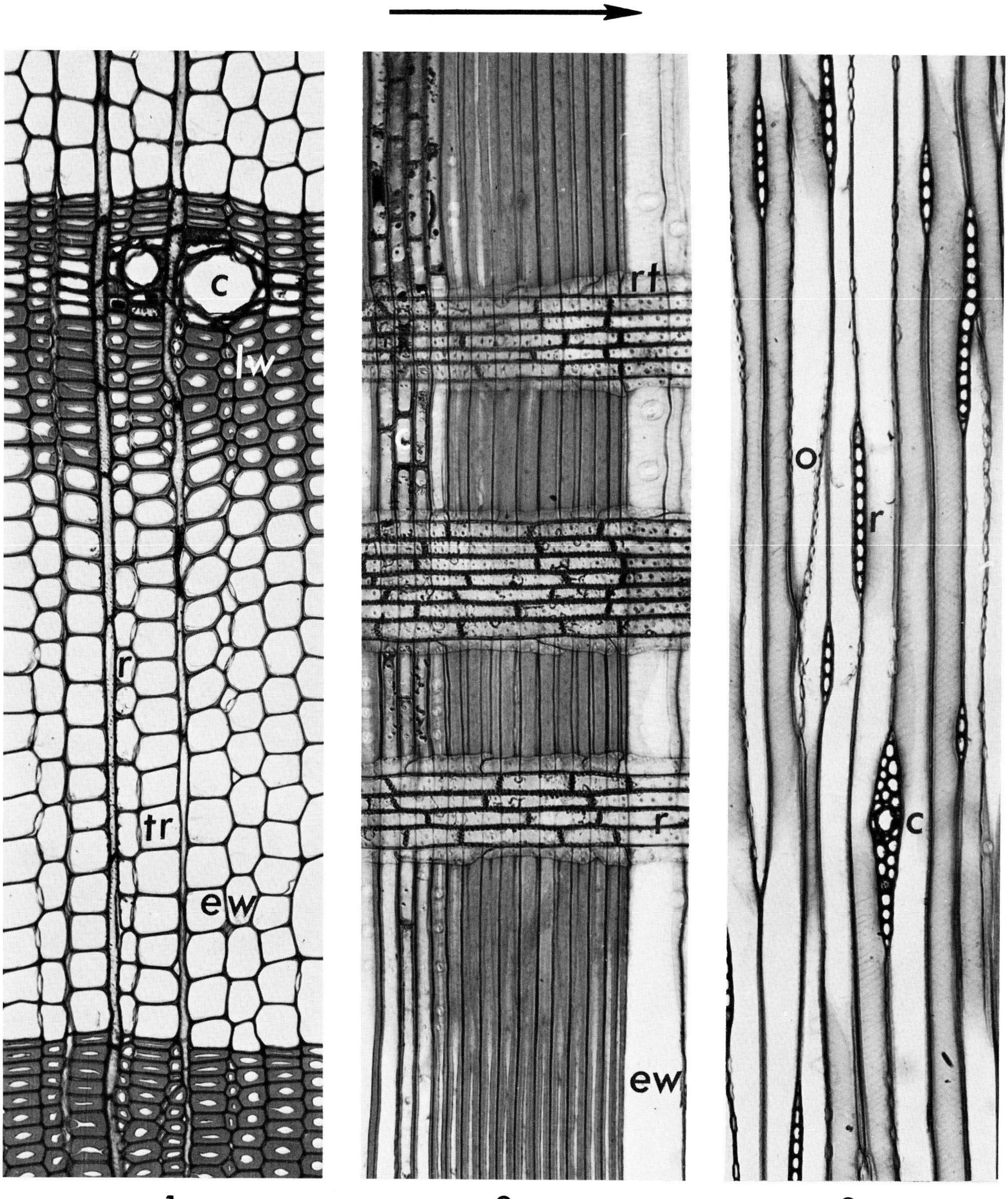

Figure 31 Douglas-fir × 175. 1. Cross section. ew, earlywood. lw (above), latewood. tr, tracheid (sap-conducting cell, about 100 times longer than it is wide). r, (left) wood ray. c, resin canal. 2. Radial section. ew, 1st earlywood tracheid of new growth. r, wood ray, rt, ray tracheid. c, resin canal lining cells. (Arrow shows direction of tree growth.) 3. Tangential section. c (left) wood ray with horizontal resin canal. r, wood ray. o, overlap zone between two tracheids (note string of connecting pits).

Softwood Micro-structure—Douglas-fir

These photomicrographs were made from stained sections less than one one-thousandth of an inch thick, cut on a machine called a microtome. A razor-sharp knife slides across the top of a small water-soaked block of wood, and the section is picked off with a camel's hair brush. Only a few people (relatively) know how to do this. One of the very first was Nehemiah Grew (see Frontispiece) who became known as "Magister Phytotomiae," Master of Plant Slicing!

The photos show for the first time in this book the truly remarkable microstructure of a piece of wood. Before studying them, go back to page 10 and get firmly in mind the structure of wood as seen in three dimensions. You are already familiar with growth rings and wood rays. Notice that in the block model, earlywood is dark and latewood is light. Negative prints were used with the thought that this gave a more "natural" look to the wood at this scale. But in the 175 × illustrations, blacks and whites are reversed giving the appearance of sections as actually viewed with a microscope.

Notice that in the cross section there is one complete growth ring. At the bottom is the latewood of the previous year (or growing season), and at the top the earlywood of the following season. The tracheid walls of the latewood are very thick and the cell cavities so small that little if any sap conduction takes place through them. Latewood gives strength to the tree. Now, see the enormous change in size and wall thickness between a last tracheid in the latewood and its following cell above in the earlywood. In spring, or whenever the tree begins to grow (consider high mountain, or desert sites) it really gets off with a sprinting start. The walls of some of the tracheids appear to be split. Later we shall see that these are remarkable little "valves" through which sap passes between tracheids. The two long lines of cells running from top to bottom are wood rays, and the two "holes" at the top are the cavities of resin canals found normally in pine, larch, spruce, and Douglas-fir, and occasionally in other conifers supposedly as a result of wounding the tree. The so-called Naval Stores industry obtains turpentine and rosin by wounding certain hard pines and collecting the resin that oozes from these canals.

Wood rays feature the radial section, three of them sweeping across the illustration. At the top and bottom of each ray is a single row of empty, dead conducting cells, the *ray tracheids* that connect with both the longitudinal tracheids and also with the other cells of the rays. These other cells making up most of the ray are alive in sapwood. They are filled with protoplasm and together serve as food storage and translocating tissue. Beside sugars and other carbohydrate foods, they may contain oils and other substances, many unknown. Running vertically along the left edge (top) of the photo is a band of rectangular cells, the lining *(epithelium)* of a resin canal. These cells secrete resin into the canal.

The tangential section shows that in Douglas-fir, the rays are one cell wide except where they contain a horizontal resin canal. Now for the first time, you can see that the tracheids have closed tapering ends, and that these overlap each other. In this overlapping zone especially, there are numerous connecting pits between tracheids through which sap flows. The remarkable structure of these pits is the next subject to explore, but before turning the page notice the spiral thickenings in the tracheids. Only a few kinds of conifers have these.

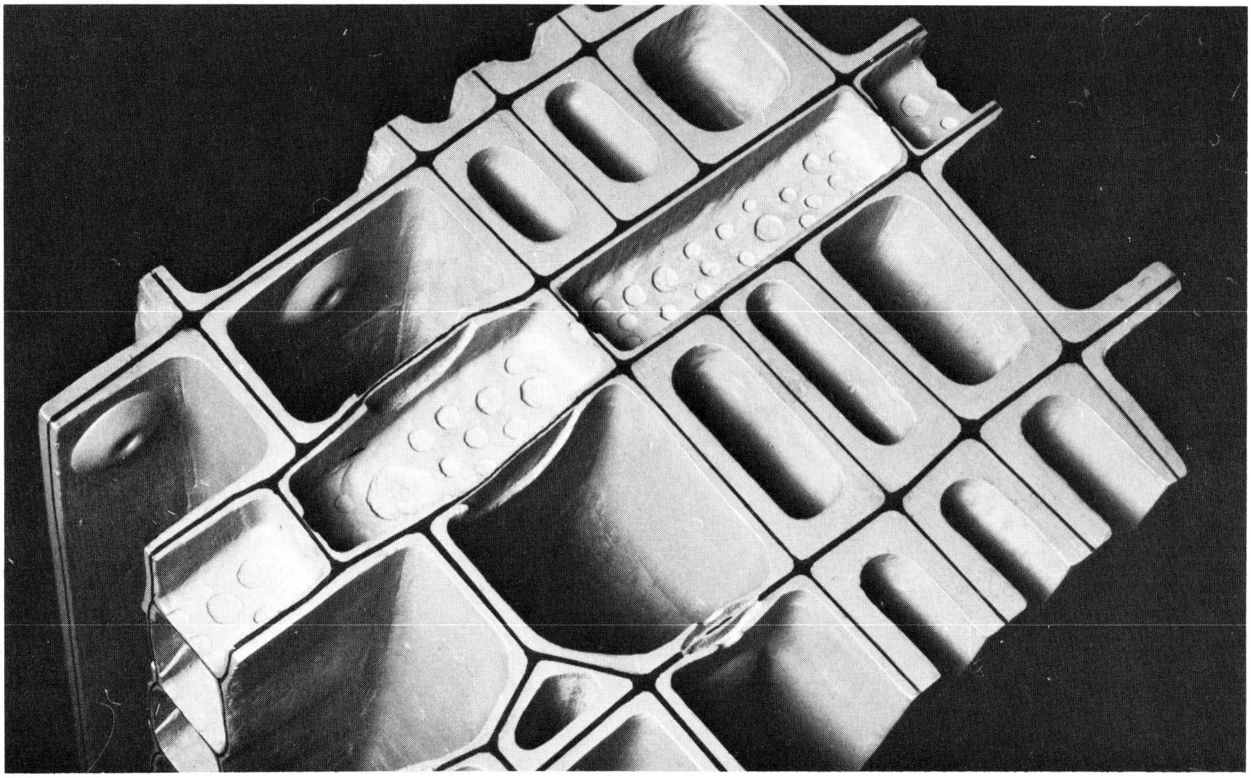

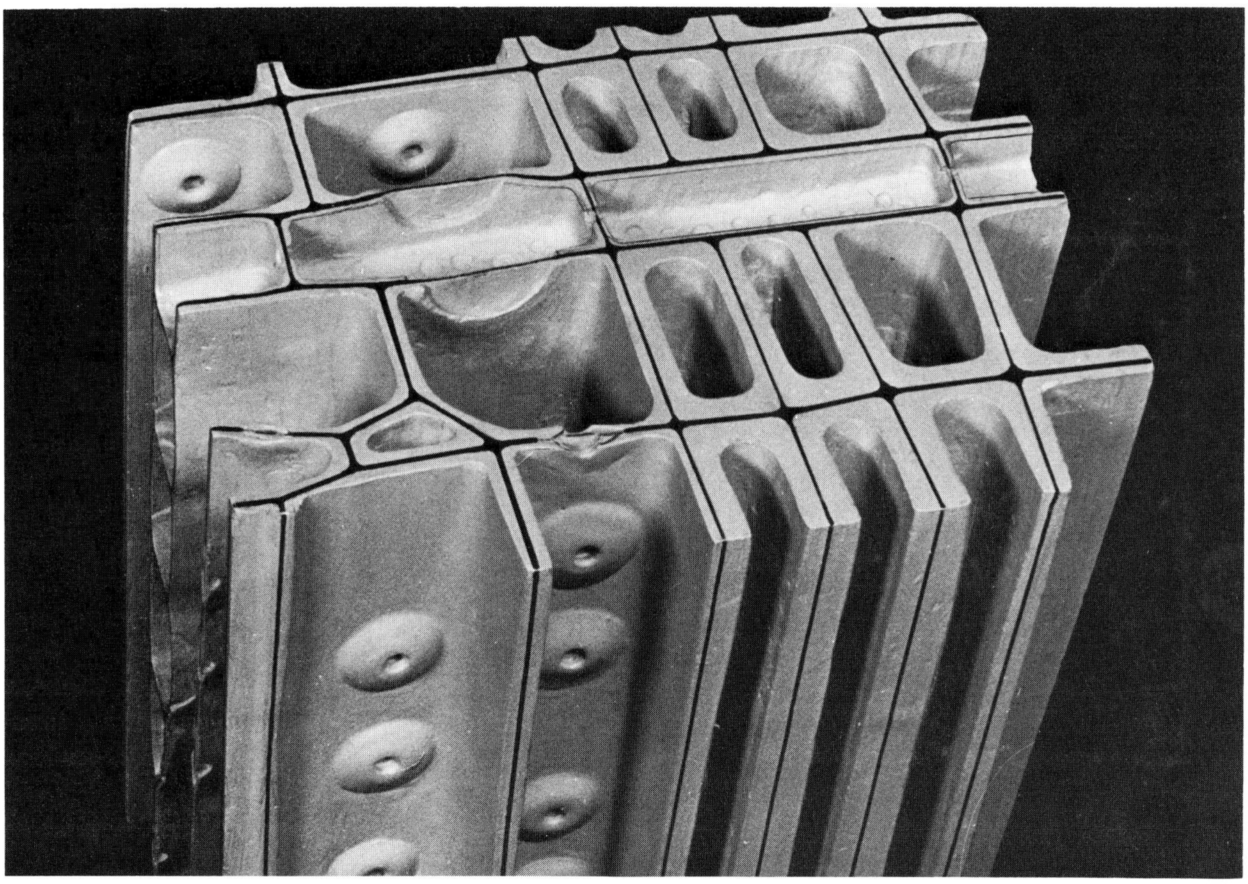

Softwoods—Sugar Pine

Here is a hand-carved three dimensional model of sugar pine wood. At this point you must make the jump from a magnification of 175 (the Douglas-fir pictures p. 28) to one of about 600 times. Compare carefully the two pages of pictures with each other, and cross-identify the structural "elements" or cells.

◀ **Figures 32 and 33 Sugar pine block model showing latewood (right), and earlywood (left) of the next growing season. Growth is from right to left. × 600.**

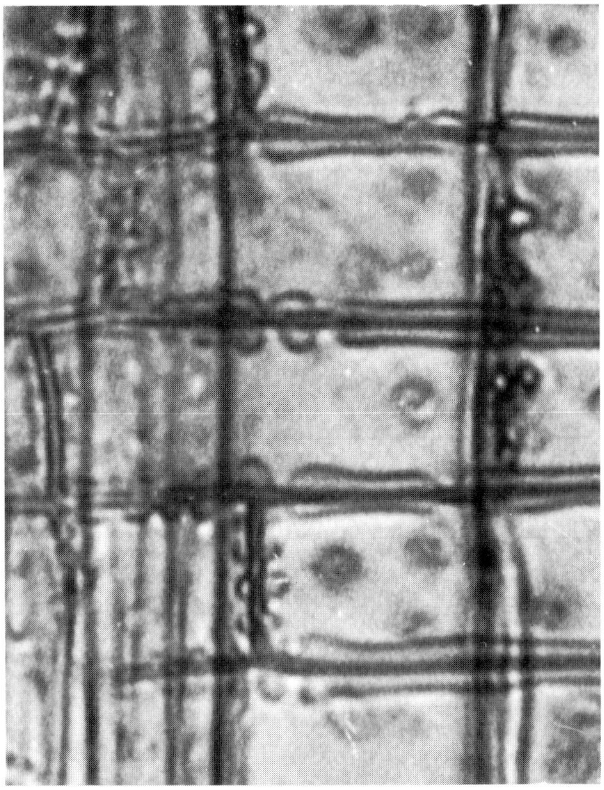

Figure 34 Balsam fir radial section showing "edge" view of simple pit pairs × 960.

Between these is the black network, the *middle lamella.* This all enveloping layer may be said to hold the cells together. If one removes it chemically, then they will tend to fall apart and this is exactly what happens during the paper pulping process (p. 111). To give an idea of size in the model, the radial dimension across the block represented is less than .2 of a millimeter, barely visible to the naked eye.

With a little imagination, you can now see that the wood in a tree trunk is the most complex, beautifully structured water transporting system in the world. Think of the forces involved in the conduction of tons of sap all the way from the smallest rootlets far underground through the larger roots and then the towering trunk of a 300 ft. tree to the branches, twigs and finally the leaves. Since the cells in softwoods are all closed units there must be some way for the sap stream to pass from one to another. This is done through the countless billions of *pits* in the cell walls. The pits allow passage of water containing dissolved minerals from the ground, and food materials made in the leaves, and stored in the roots.

Pits are holes, and some of these can be seen in face view in the wood ray cells of the model at the top of the facing page. (Due to a peculiarity of human eyesight, after you look at these depressions for a few seconds, they will suddenly look like "bumps". Keep looking and soon they will become depressions again.) In a radial section (balsam fir) you can see that the holes penetrate the outer cell wall layers but not the middle lamella. Usually a pit on one side is matched by a pit on the other, and together they are called a *pit pair.* These *simple* pit pairs are found in living cells throughout the tree.

In the dead conducting cells (tracheids) which comprise most of the wood, there is found another type of pitting. The *bordered pit pairs* are most remarkable structures. In face view on the radial surface of the model you can see seven of the pits looking like shallow saucers with a hole in the center. Above the row of four is shown in cross section another one, but a higher magnification is needed to see the structure in detail.

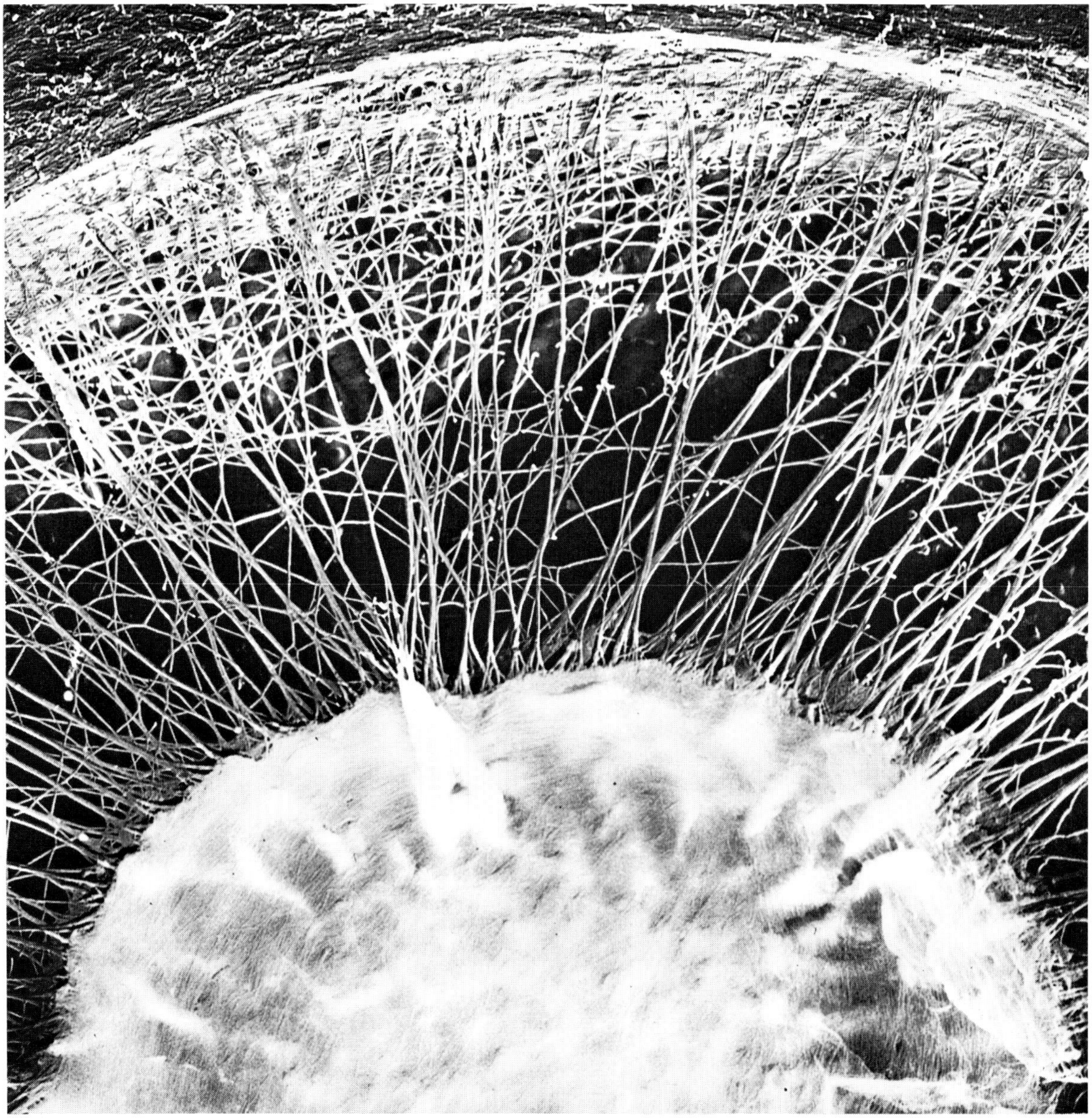

Figures 36 and 37 Bordered pit pair models × 2,400

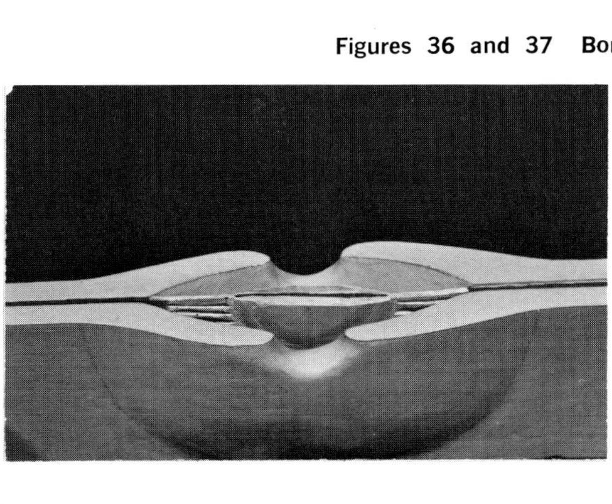

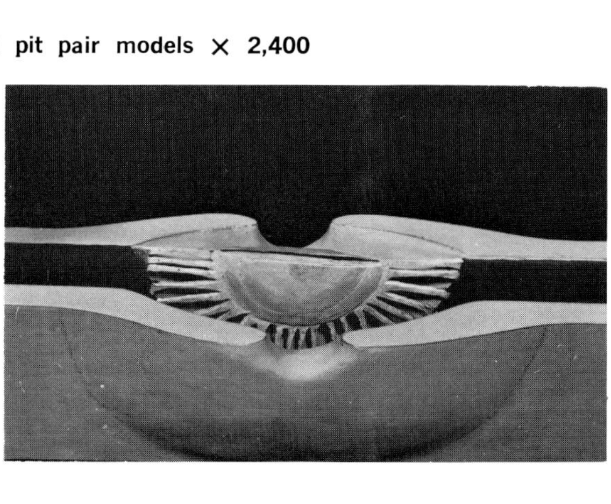

◀ **Figure 35 Electron micrograph of one half of the torus, showing the connecting strands × 16,000. (Courtesy W. A. Côté, Jr. and A. C. Day.)**

The superb electron micrograph filling most of the opposite page will have more meaning if you will first study the bordered pit pair model below.

This model was carved to represent a pit pair in cross sectional view through the center. You can see the "half saucers", each in an adjacent tracheid, the "half holes", and the most remarkable structure of all, the *torus,* the thickened membrane in the center attached by strands of microfibrils to the inside of the pit pair cavity. The illustration on the right shows the secondary wall pulled away so you can get a better view of the torus. If you construct in your imagination the complete torus you will have a circular body attached all around with its strands to the inside of the pit pair cavity.

There is enough flexibility in the strands so that under positive or negative pressure in adjacent tracheids, the torus could move to one side or the other completely sealing the pit and stopping the flow of sap through it. Does this happen in the conducting sapwood? We do not yet know. Usually, when sapwood becomes heartwood, the torus will be found on one side or the other, thus sealing the pit. This is one reason why heartwood of most conifers is usually more difficult to impregnate with chemical solutions than is sapwood.

The electron micrograph above is nearly seven times the size of the model and shows the beautiful structure of the strands as they really are. It seems clear that sap should have no difficulty in passing through this coarse network from tracheid to tracheid.

The torus is not present, so far as known, in bordered pit pairs of the hardwoods. Their enormous vessel tubes carry most of the sap, and very little of it passes through the fibrous elements.

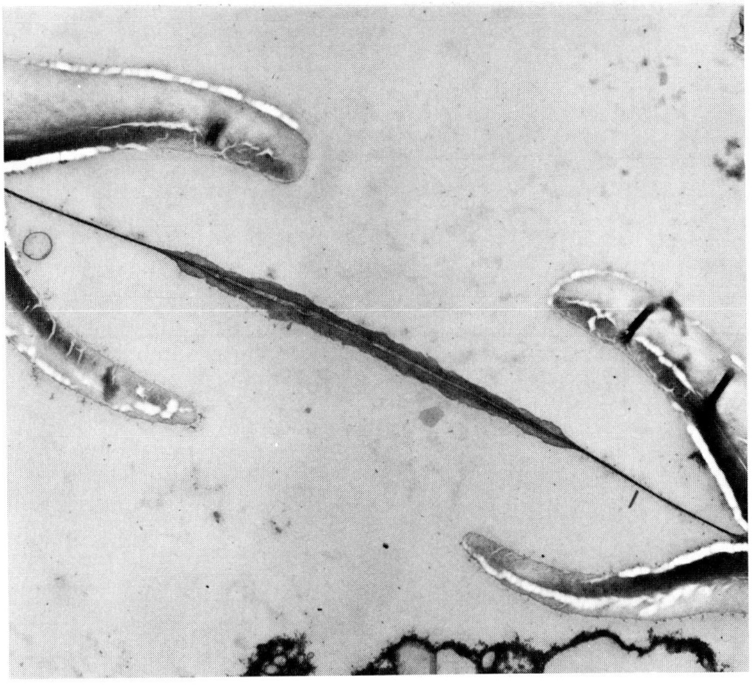

Figure 38 Electron micrograph of a bordered pit pair showing the torus × 7150. (Courtesy of Norman P. Kutscha)

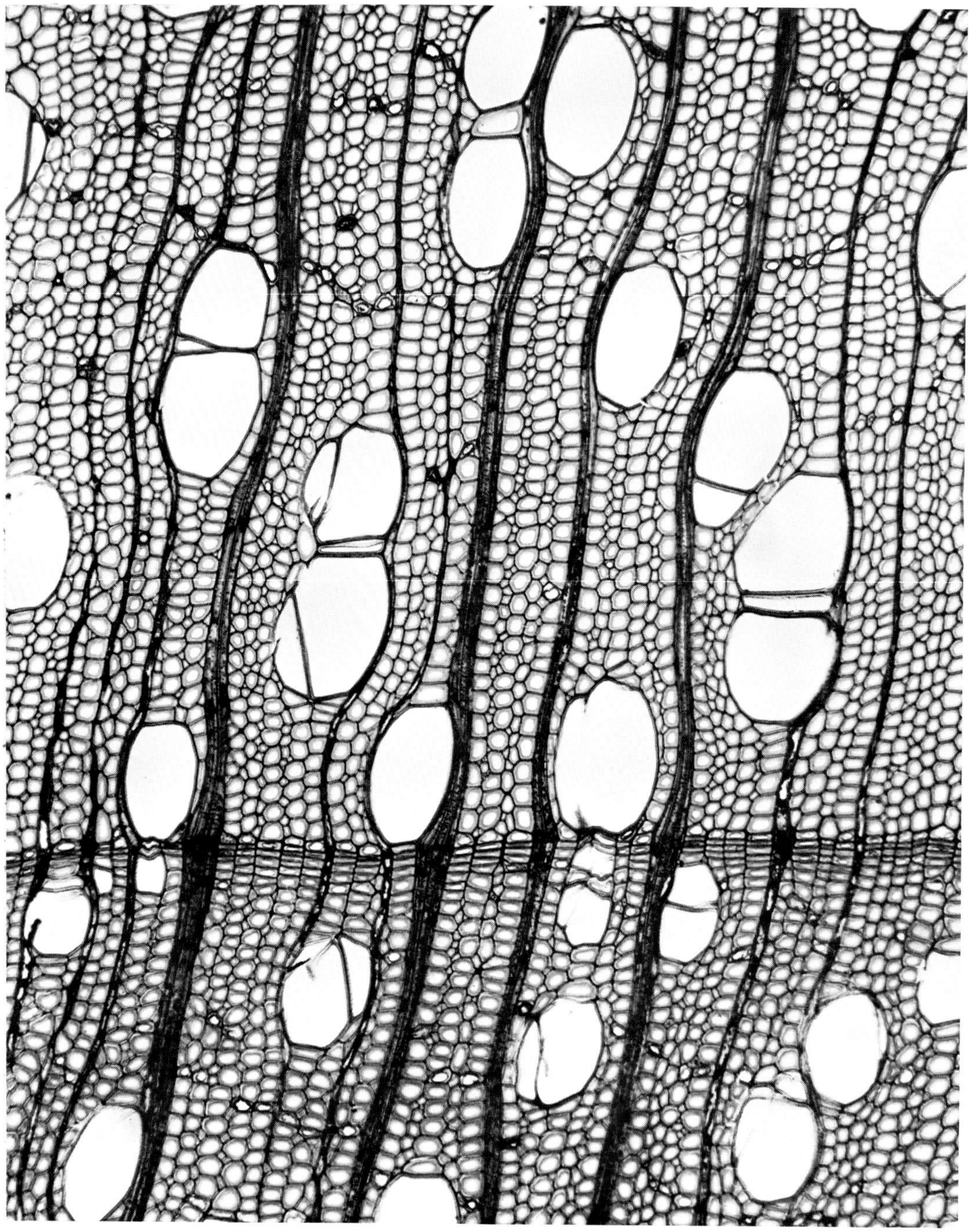

Figure 39 Yellow birch cross section × 175. Latewood is at the bottom, and above it the earlywood of the next season.

Hardwood Micro-structure

The wood of yellow birch like that of other broad-leaved trees shows a great difference between the size of the sap conducting cells and those that serve principally to strengthen the tree. The immense *vessels* (their openings are *pores*) allow large volumes of sap to pass through quickly without having to "filter" through myriads of small pits. Besides the wood rays which may be from one to several cells wide, most of the other cells are *fiber tracheids* with thick walls, having bordered pits. Scattered

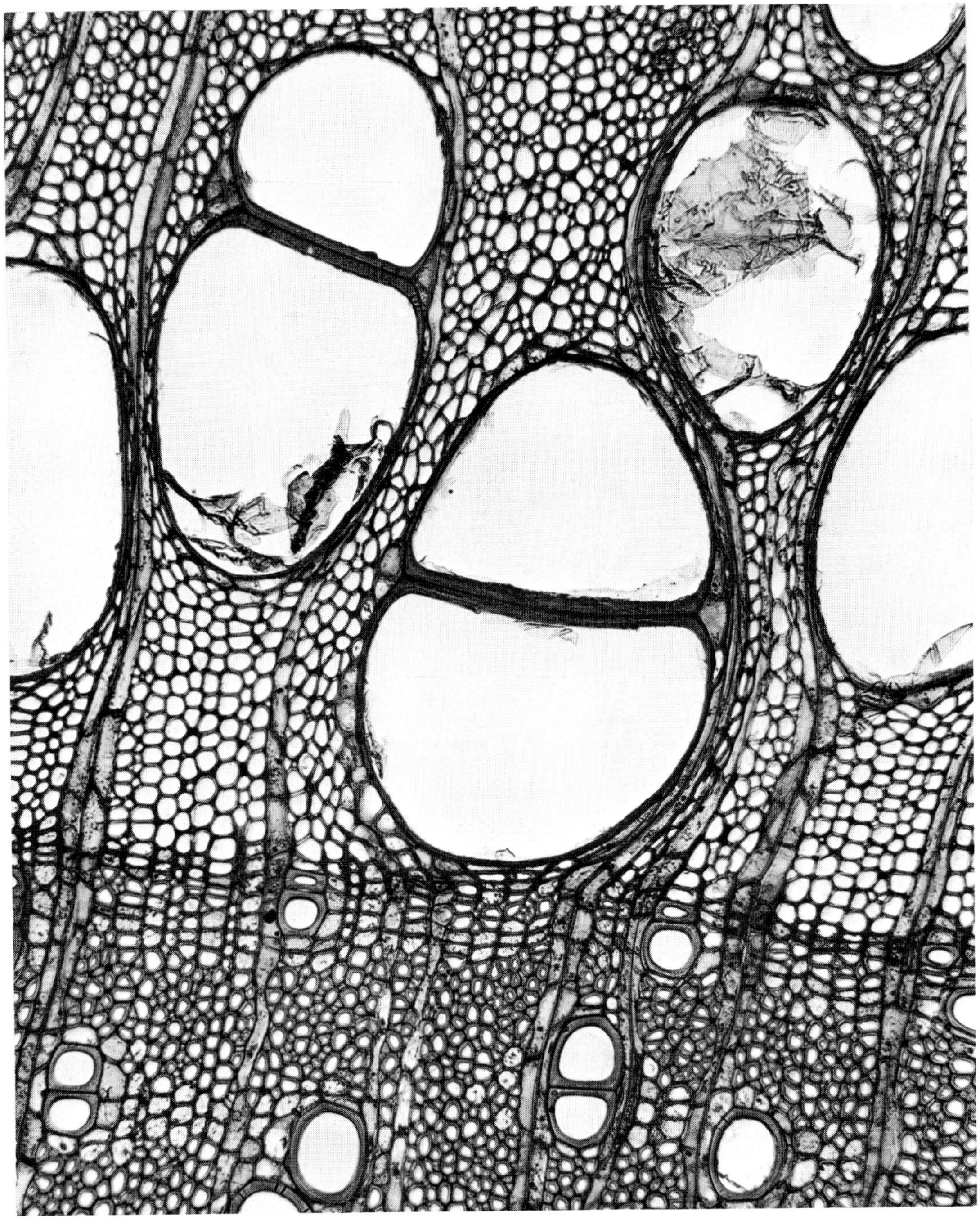

Figure 40 White ash cross section × 175, latewood at the bottom, earlywood above it.

throughout the section are thin-walled live cells *(parenchyma)* whose contents may be black in the photo.

The wood of white ash shows still greater specialization between the vessels and the other cells. In the earlywood, the thin-walled woody cells surrounding the enormous vessels are tracheids. The much smaller vessels of the previous season's latewood below are more or less connected by bands of thin-walled parenchyma, and above and below these are groups of very thick-walled cells, the *libriform fibers* with simple pits. As in some other woods, there are often bag-like protrusions which grow from a live cell next to a vessel into its cavity. Parts of these *tyloses* show in the vessels.

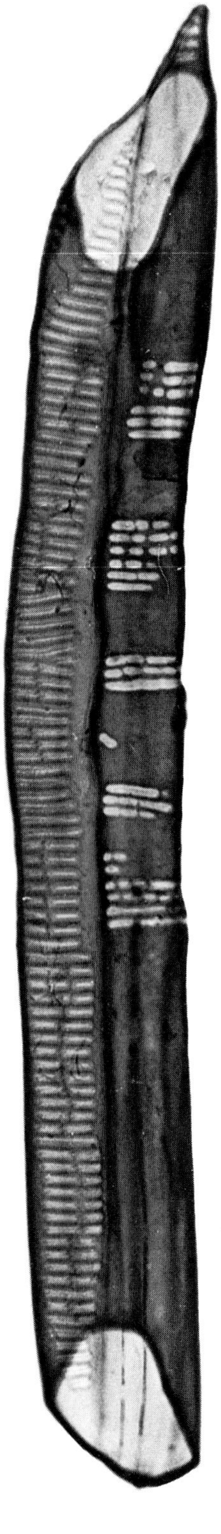

Figure 41 Magnolia vessel segment × 200

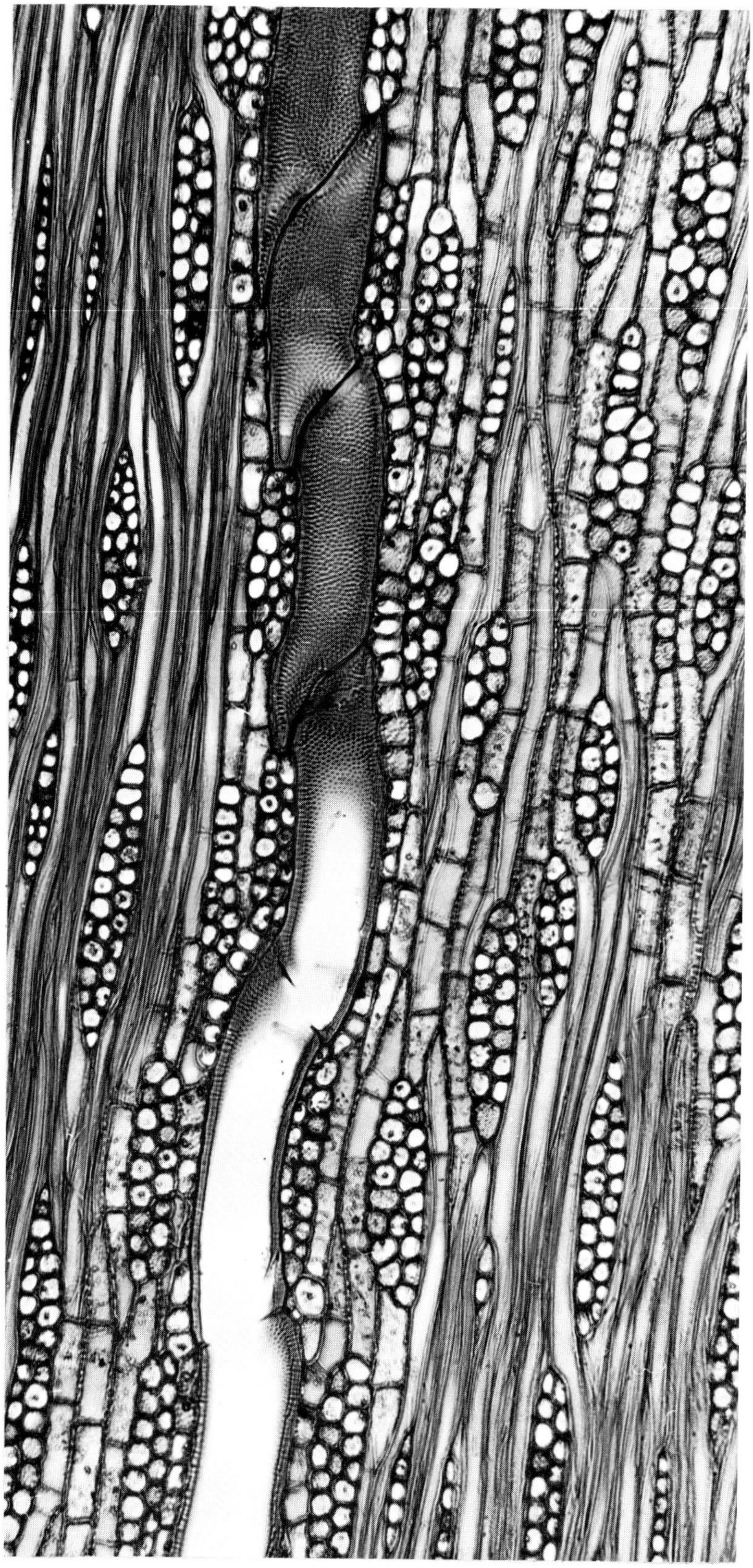

Figure 42 White ash tangential section showing a chain of vessel segments in the latewood × 300.

The Hardwood Vessel

Because of their immense size, the earlywood vessels in ring-porous woods are perhaps the most remarkable structures found in wood. Turn back to page 35 and observe that the wood ray at the right of the largest pair of vessel openings curves around them and then reverses its curve around the vessel above containing tyloses. It looks as though the rays had been pushed aside by the vessel during its growth, and this indeed is what happens. The vessels originated from cells no bigger than the small tracheids which surround them. Growth enlargement in diameter is about 20 times. Something has to give, and not only the rays but also the tracheids are squeezed, and some flattened.

The white ash tangential section (facing page) shows a vessel tube extending from top to bottom. You can see that the tube is made up of six short *vessel segments*. The wood section happened to be cut so that the outer heavily pitted walls of the top three segments are exposed. The vessel may have had a slight curve, and the knife sliced through the lower three segments showing them in edge view. Between nos. 4 and 5, and 5 and 6, notice the small "spines" that mark the ends of these vessel segments. At some point during the enlargement of the vessel from a vertical string of small, growing cells, the end walls of these cells disappeared by a process still not entirely understood. This left a long clear tube with no interruptions except the "spines" which mark the ends of the cells. Notice that the openings of the magnolia vessel are also clear of obstructions. In some trees (yellow-poplar, birch, and others) instead of clear *perforations* between vessel segments, one sees a number of narrow horizontal parallel bars across the perforations.

It has always been wondered how far up and down a tree an uninterrupted series of vessel segments might extend. It has been shown that eventually there is a tapering "dead end". Here the sap stream must flow laterally into an adjacent vessel or through connecting tracheids or fiber tracheids into a vessel that presumably continues the flow upwards. The continuous tubes are much longer in the ring porous woods than in the diffuse porous ones. If you take a 1″ x 1″ piece of red oak wood (ring porous) two feet long (probably even longer) and dip one end in a soap or detergent solution, you can then blow soap bubbles with it! Experiments with ash wood have shown that coal gas can be forced through a branch 10 feet long and ignited on the other end. With maple (diffuse porous), the length was only 2 feet, and with a softwood, even a piece 1½″ long stopped the flow of gas.

The separate vessel segment of the magnolia was photographed in an unusual position so that it shows bordered inter-vessel pitting (left), and pits that connected with wood rays (right). When I first mounted it in Canada balsam on a slide with a cover glass, and looked at it with a microscope I was delighted to see that two sides were visible. Almost always these things roll over one way or the other and show either inter-vessel, or ray pitting, but not both. I put the "wet" slide away very carefully. Next morning I looked at it. The vessel segment had rolled over and now showed only one side. I barely touched the edge of the cover glass with a dissecting needle and rolled the segment back again. This was a very delicate procedure; the vessel segment is only about one millimeter long. The following morning and the next, adjustment was still needed until finally the balsam resin dried out enough to hold the vessel just as you see it.

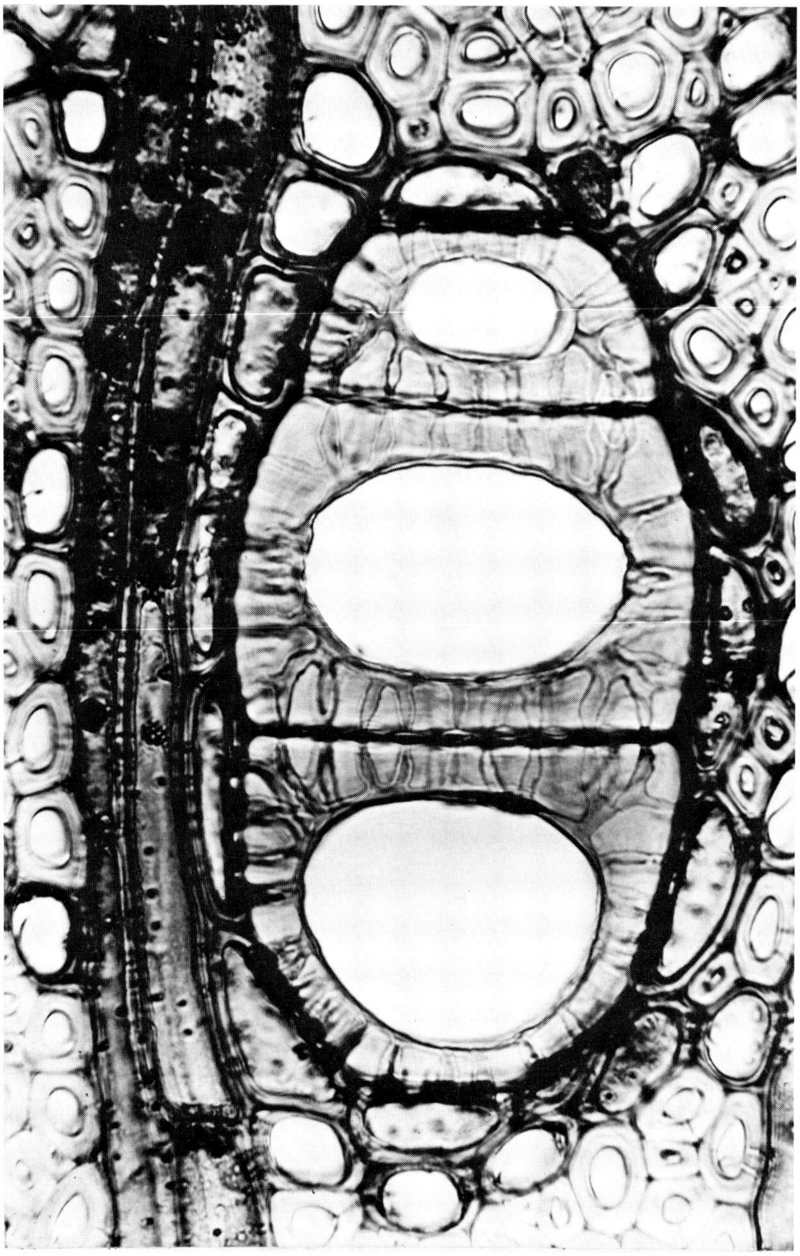

Figure 43 Hickory cross section of latewood vessels showing interconnecting pits × 750.

The unit of measurement with the light microscope is the micron (μ). One millimeter (approximately 1/25 of an inch) equals 1,000 microns. In the picture above, the inside horizontal diameter of the center vessel is 44 mm. Divided by the magnification (750), the actual size is .0587 mm or 58.7 microns.

Chapter III

A Look At Some Common Woods

These are only "common" in the sense that they are widely used, and available in quantity. Actually, each one is an exquisitely beautiful structure having features which separate it from other kinds of wood.

With some practice, you can cut thin sections of wood with a single edged razor blade. Small blocks about ⅝″ along the grain and ⅜″ square in cross section are a convenient size. Boil them in water until they sink to the bottom. This may take several hours. At this point, the lighter, softer ones may be sectioned and viewed with a low power microscope. The heavier, harder woods require two or three days to a week of boiling before they are soft enough to cut. Boiling can be done in any sort of pan, or glass beaker, but don't forget to keep plenty of water over the blocks, or they will go dry and burn. You then become automatically a member of the "wood burner's club". All this can be avoided in the laboratory by using a flask, topped by a water cooled reflux condenser.

The cutting technique is important. Hold the block between thumb and forefinger of one hand, brace your elbows against your body and, with the razor blade, make diagonal slicing strokes across the top of the block. Some people learn how to do this quickly, others take *much* longer. It is usually not possible to cut a good section from the entire top of the block, but only a partial one will show structural features. The purpose is to get as thin a section as possible otherwise you won't see much detail with the microscope. Professionals, using a special machine (microtome) cut sections from 5 to 15 microns in thickness.

"Textbook of Wood Technology" (8) lists and describes some 75 important kinds of wood that grow in the United States. Actually, there are between eight and nine hundred tree species in our country, but only a relative few are of commercial value. Also, in certain groups it is not possible to separate the species from each other by wood structure. For instance there are about 60 oak species in the United States. They can be readily separated by wood features into the white oaks and the red oaks (see p. 46). But further separation within each group cannot be done with certainty using the structural features of wood. There are other groups such as the southeastern hard pines in which species cannot be separated by wood features.

Twenty-seven species or groups have been selected for this book, not only for their importance but also to show the variation in structural features. The specific gravity numbers are averages based on oven-dry weight and green volume. See p. 70 for explanation of the method.

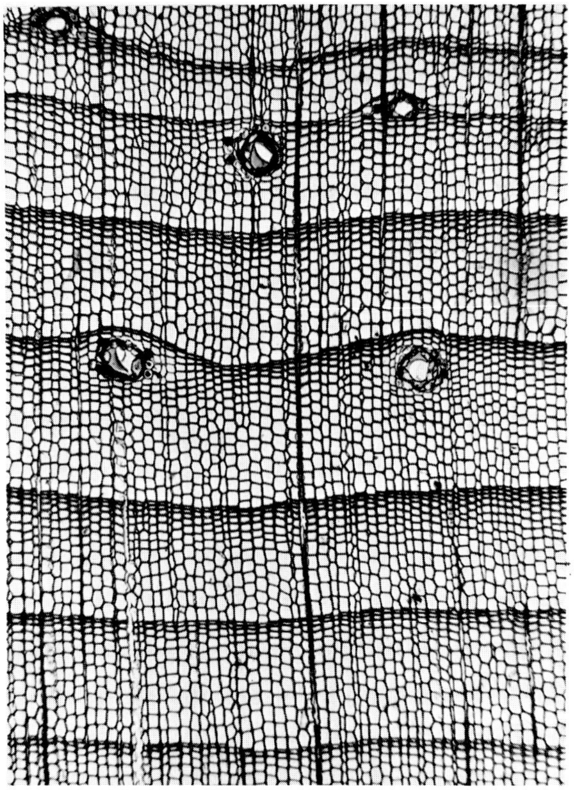

Figure 44 Eastern white pine × 25 Sp. gr. .34

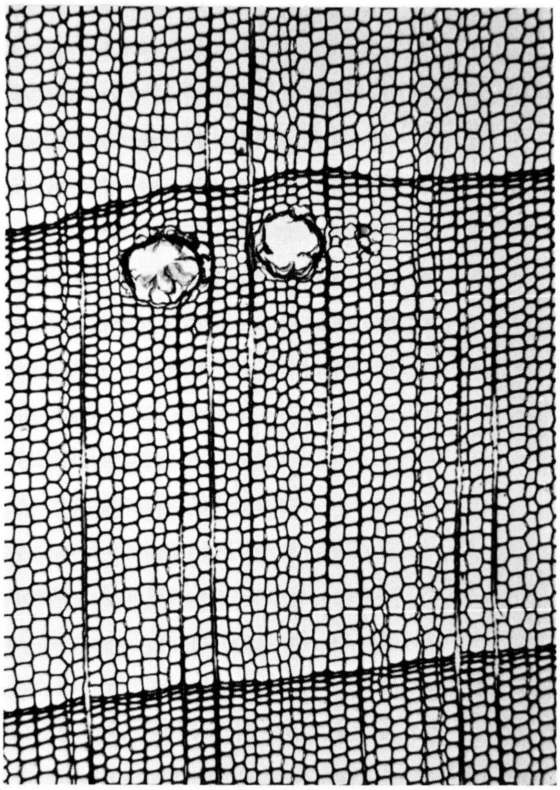

Figure 45 Sugar pine × 25 Sp. gr. .35

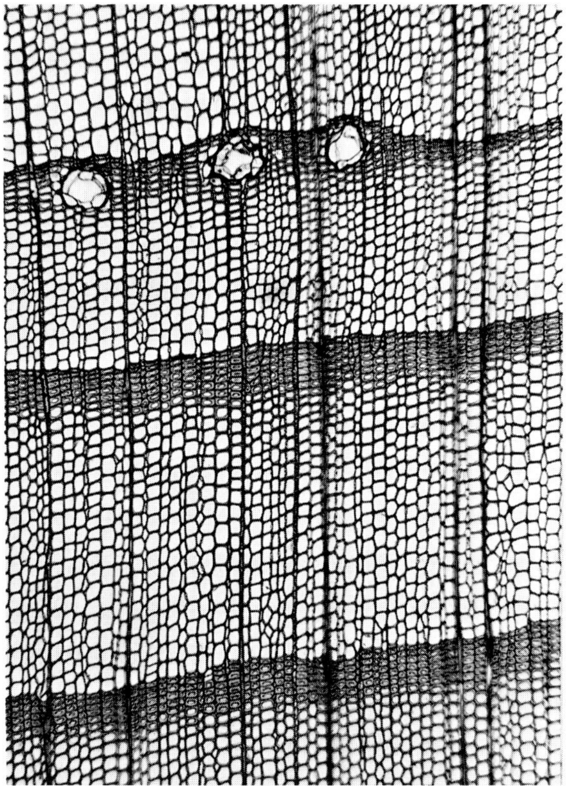

Figure 46 Ponderosa pine × 25 Sp. gr. .38

Figure 47 Slash pine × 25 Sp. gr. .56

Pines are mostly very important timber species. There are about 35 of them in the United States, over much of the country except the plains and prairies. Pine, larch, spruce, and Douglas-fir all have longitudinal resin canals, but they are usually most easily seen in pine. On the opposite page are wood cross sections of four of our most important pines. Before you note their specific gravities, make your own estimate of their relative weights based on the amount of latewood in each species.

Eastern White Pine was the monarch of all conifers during the colonial period of our history. It towered over all other of its associates, with a straight trunk sometimes 200 feet tall and 4 to 6 feet in diameter, clear of side branches for two thirds of its length. A magnificent tree it was, and the supply seemed inexhaustible, first in New England, then in New York, and Pennsylvania, and finally in the Lake States where vast stands once covered the land. They are nearly all gone (except in parks) but plenty of new growth is coming along. At first whole houses were built of this pine with its soft, easily worked wood. But now with lumber cut from small trees with many side branches, the knotty wood is largely used for *boxes* and *crates*. Recently, however, the beauty of knotty pine for *interior panelling* has been discovered. Clear white pine is favored for *window sashes, doors,* and many other things. One outstanding use is for *foundry patterns*. These are pushed into special sand to make an impression or mold into which molten metal is poured. Since the metal shrinks upon cooling, the wood pattern must allow for this, and the carving and finishing of patterns is a highly technical art. In the Northwest, a very similar tree, western or Idaho white pine, is found.

Sugar Pine of California is the "king" of all pines with a massive cylindrical trunk, sometimes well over 200 feet tall and 6 to 10 feet in diameter. The wood although similar to that of the other two white pines, has a slightly coarser texture. Used as are the other white pines, two unique additional uses are for *piano keys* and *organ pipes*. The name of the tree comes from a sugary substance that exudes from trunk wounds.

Ponderosa Pine is the outstanding pine of our western states. Notice that although the transition from early to latewood is more abrupt, and there is more of the latter than in the white pines, this hard pine (yellow pine) looks more like a white pine (soft pine) than it does the slash pine representative of the Southern hard pines, with its very heavy band of latewood. Ponderosa pine is also softer than the Southern pines. For this reason millions of feet of Ponderosa pine (western yellow pine) have been sold and used as "white pine." Besides these uses, Ponderosa pine is the "building wood" *(joists, rafters, studs, sheathing, interior trim, poles* and *posts*).

Slash Pine (Southern yellow pines). Longleaf, shortleaf, loblolly and several other pines are included since they are not separated by wood structure. The band of heavy latewood features these woods. Used in many ways, *heavy timbers, railroad cars, ships, paving blocks,* and *pulpwood* from vast stands of young growth should be mentioned. Also since the beginning, slash and longleaf pines have furnished the valuable *"naval stores"*, turpentine and rosin. The trees are wounded and the exuding resin is collected and steam distilled. The volatile part is turpentine, the solid cake remaining is rosin. Perhaps you only know rosin as used to stroke violin bows, but tons of powdered rosin are used in making paper, and for other industrial uses.

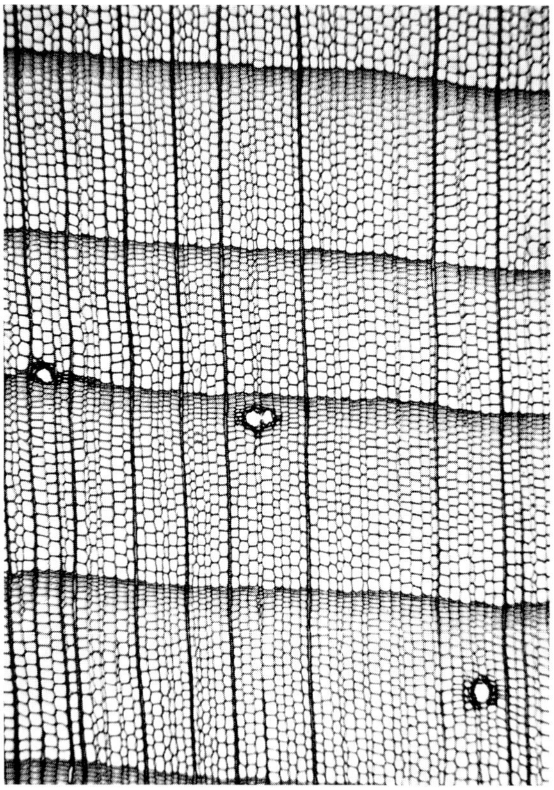

Figure 48 Douglas-fir × 25 Sp. gr. .48

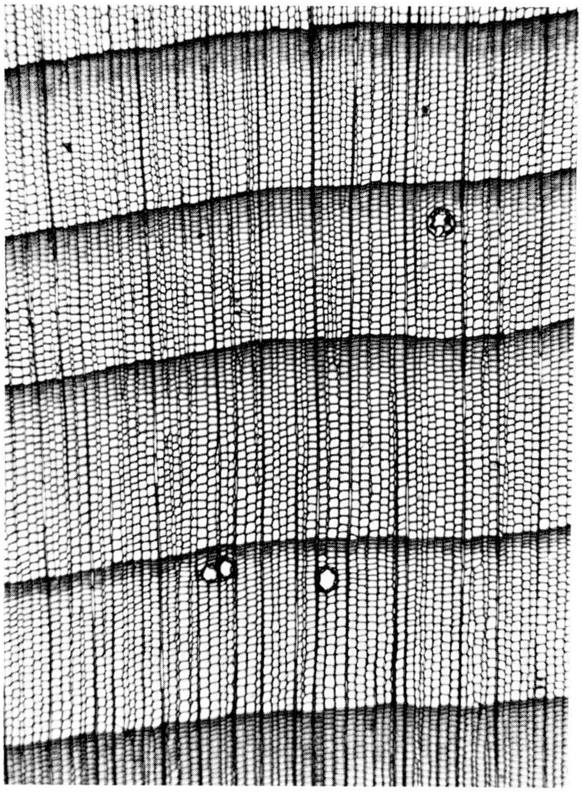

Figure 49 Red spruce × 25 Sp. gr. .41

Douglas-fir furnishes more timber than any other single species in the United States, and our western forests are about 50% Douglas-fir. It is one of the forest giants, and old trees may be 250 feet tall or more, and 8 or more feet in diameter. Single trees may rival the redwood.

The wood is of moderate weight and the heartwood is variable in color ranging from pale yellowish red through orange to deep red or reddish brown. The resin canals are smaller than those of most pines, but are still visible with a hand lens. Perhaps the best feature is a microscopic one seen on p. 28. The conspicuous spiral thickenings in the tracheids are found in no other coniferous woods of North America with the exception of yew, and torreya, but these woods lack resin canals.

The wood of Douglas-fir is widely used for *building* and *construction,* including *veneer* fabricated into *plywood; railroad ties* (usually treated with creosote to resist decay), *ship* and *boat building, furniture, mine timbers, wood pulp,* and others.

Red spruce, an eastern tree represents the several spruces found in both eastern and western United States, and in a vast forest across Canada. The two pictures look similar, but spruce wood is lighter both in color and weight than that of Douglas-fir. The latewood in Douglas-fir is usually wider and more conspicuous than in the spruces, but remember there is great structural variation in wood, and these two pictures furnish an example.

The wood of spruces is pale yellowish brown to nearly white, and sapwood and heartwood show no distinct differences in color. The most important use is for *pulpwood* followed by many others including *sounding boards* (the wood has high resonance) for *pianos, boxes* and *crates ladder rails, paddles,* and *oars* (spruce wood is fairly strong, but also stiff, and light), and *planing-mill* products of all sorts.

Redwood of the West Coast is the tallest tree in the world (about 370 ft. maximum). Diameters are from 8 to 12 ft. but may reach 20 ft.

The heartwood varies from light to deep brownish red, and because of chemicals naturally impregnating it, this wood is unbelievably resistant to decay. Fallen trees after lying on the ground for 1,000 years still have sound heartwood. Redwood is one of the most coarse textured of the conifers (see photo).

The uses include *building construction* of all sorts, *tanks* and *vats, conduits* for carrying liquids (the wood has high resistance to chemicals, does not warp or twist, and has low shrinkage); *ship* and *boat building, caskets* and *coffins, shingles* and *shakes,* and many others.

Port-Orford-Cedar represents a number of American trees called "cedars", but the true cedars belong in a group of Old World species *(Cedrus)* of which the Cedar of Lebanon is the most famous.

The native cedars have fine textured woods each of which has its own characteristic aromatic odor. Like the redwood, the heartwood of cedars contains certain chemicals that are poisonous to wood destroying fungi.

Port-Orford-Cedar, a large tree of limited range in coastal California and Oregon produces wood having a strong ginger-like odor. This is the principal wood used for *storage battery separators* because of its resistance to acids, and high electrical resistance. Other uses include *boat decking, woodenware, Venetian blind slats,* the *lining* for *moth repellent chests* and *closets,* and general *millwork.*

Figure 50 Redwood × 25 Sp. gr. .38

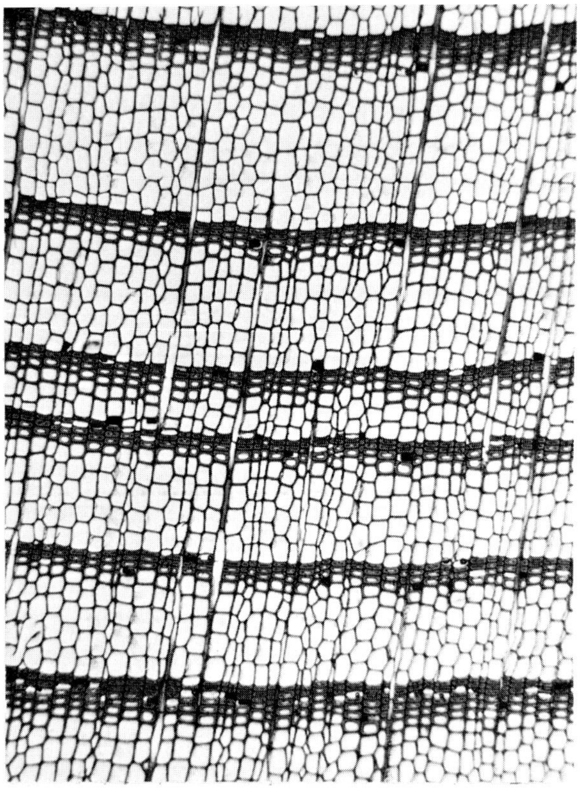

Figure 51 Port-Orford-cedar × 25 Sp. gr. .41

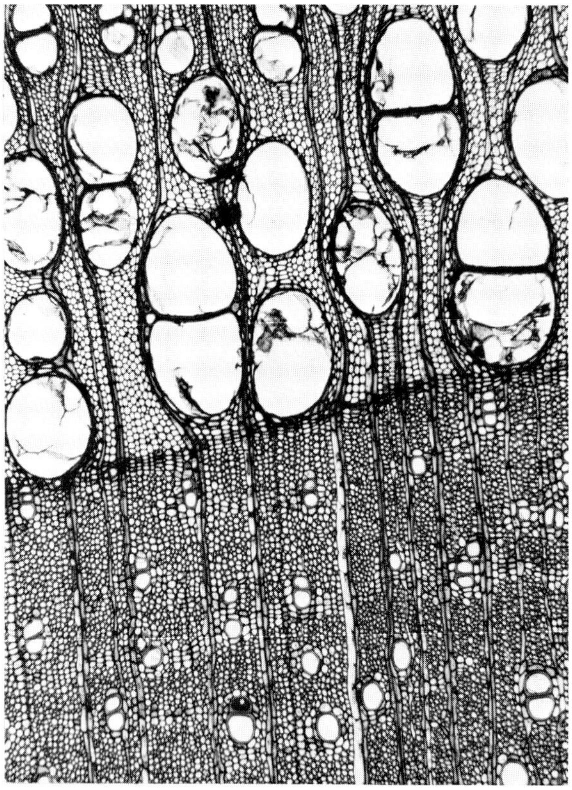

Figure 52 White ash × 25 Sp. gr. .55

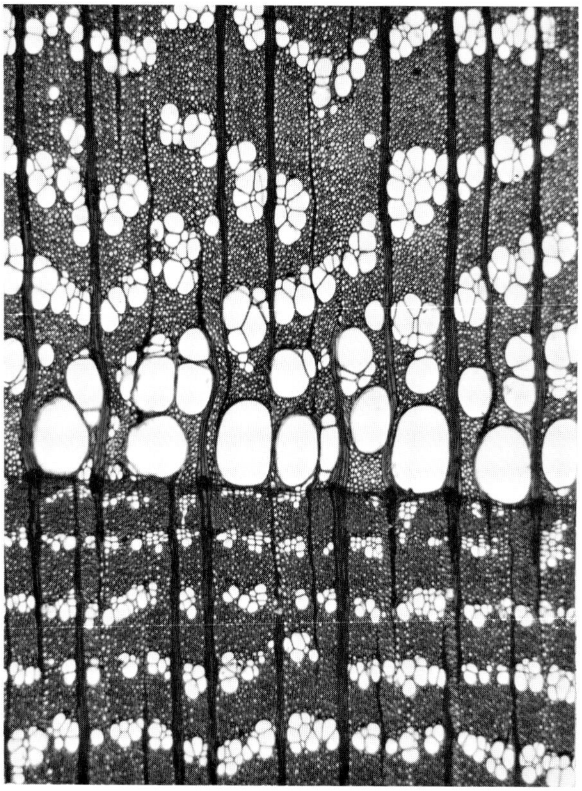

Figure 53 American elm × 25 Sp. gr. .50

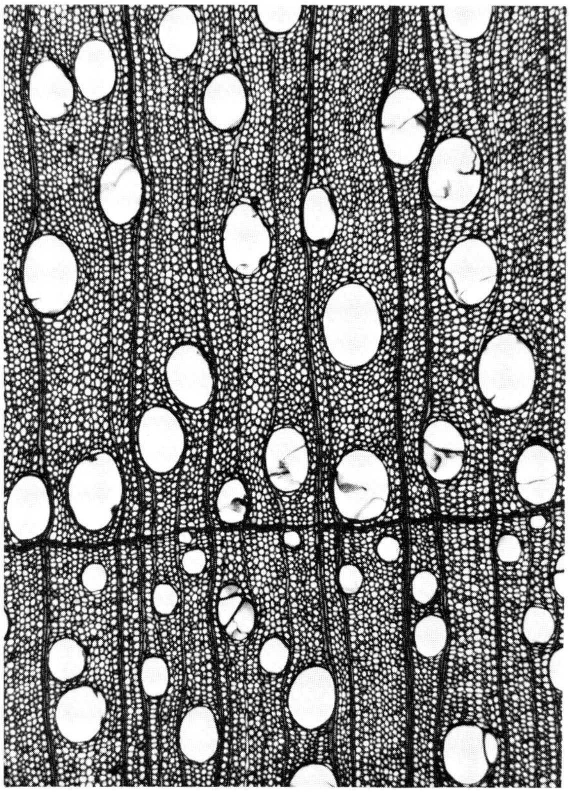

Figure 54 Black walnut × 25 Sp. gr. .55

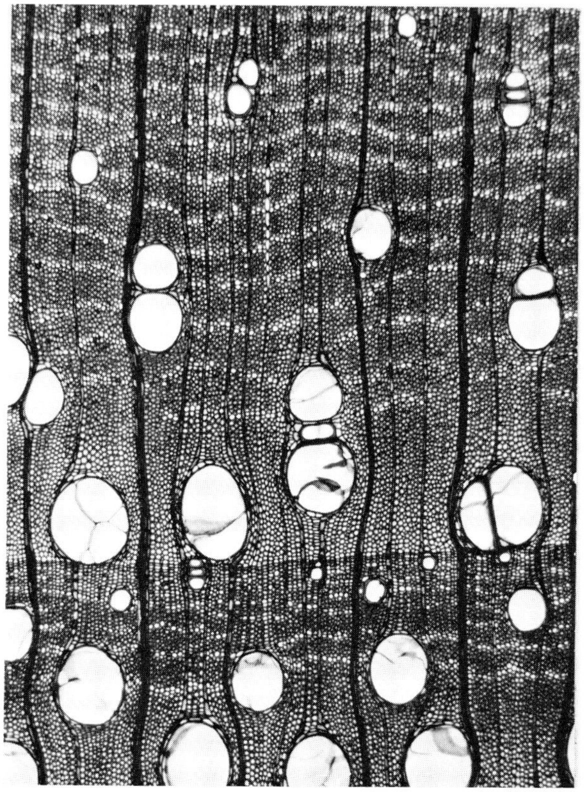

Figure 55 Bitternut hickory × 25 Sp. gr. .66

Each of these four woods comes from one to several separate species of eastern North America.

White ash heartwood is light brown in color, and somewhat shiny on a smooth surface. The illustration shows latewood below, earlywood above of a fast growing piece having larger pores than are usual.

Ash wood is outstanding for *handles* of all kinds, especially long ones (rakes, hoes, shovels, etc.) because of its springiness. This feature cushions the body from sudden jolts as one uses the tool in rough ground. Almost all *baseball bats* are of ash; other uses include *oars, paddles, snowshoes, tennis rackets, refrigerators, furniture, toys* and *woodenware*.

American elm appears to be joining American chestnut on the way to extinction. The Dutch elm disease fungus (carried by bark beetles) chokes the water conducting vessels and the tree dies for lack of water.

The heartwood is brownish, sometimes with a reddish cast; elm wood has interlocked grain and is extremely difficult to split. The uses include *boxes* and *crates* that must withstand rough handling where strength, toughness, and bending are important, *veneer* for such things as *fruit containers* and round *cheese boxes; bent parts* in *furniture;* and *vehicles*.

Black walnut has the distinction of being the finest cabinet wood of the United States, and the best grades are always in short supply. Notice that the large earlywood pores are not as localized at the beginning of the season's growth as in the two preceding ring porous woods. This half way stage toward a diffuse porous pattern is called *semi-ring porous*.

The heartwood is light brown to deep chocolate in color. Uses include *veneer* for plywood used as *paneling,* in *furniture,* and *cabinet work* of many kinds. The lumber goes directly into high grade furniture including *dining room tables* and *chairs* that must withstand hard usage. Black walnut is the leading wood for *gunstocks* because of its resistance to shock, fine machining and finishing properties. Much of this wood also goes into *caskets* and *coffins*.

Because of its scarcity, planting programs are being encouraged for black walnut. Best growth is made only on moist, deep, fertile soils.

Bitternut hickory is another semi-ring porous wood. Certain other hickories are ring porous. Hickory heartwood is brown to reddish brown, and is noted for its hardness, toughness, and resistance to shock. It is also springy and makes a good *bow*. President Andrew Jackson's nickname, "Old Hickory" was a tribute to the qualities of this wood.

Again as in ash, *handles* lead the list of uses, especially those such as *axe helves, pickaxe,* and *sledge hammer* handles and others that must stand up under repeated heavy shock. Other uses include *ladder rungs, furniture, skis, dowel rods,* and *skewers*. Hickory is a prime *fuel wood,* and is also famous for *smoking meats*.

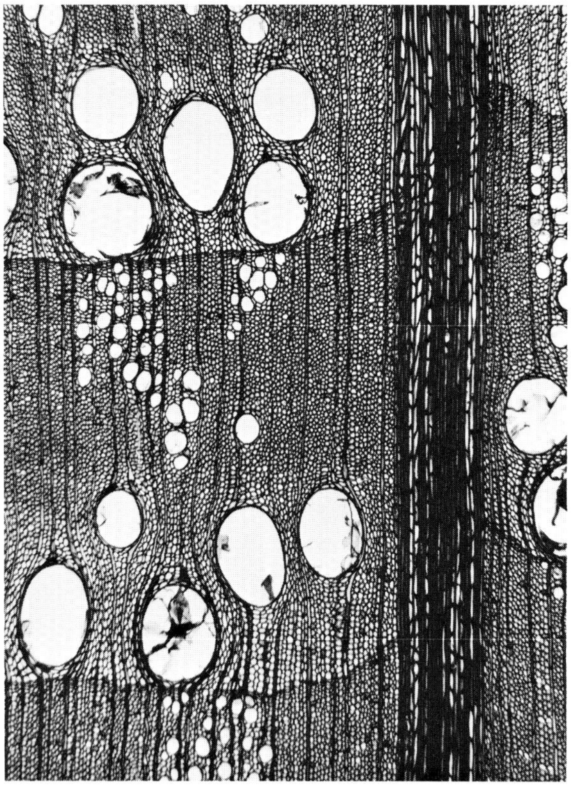

Figure 56 White oak × 25 Sp. gr. .60

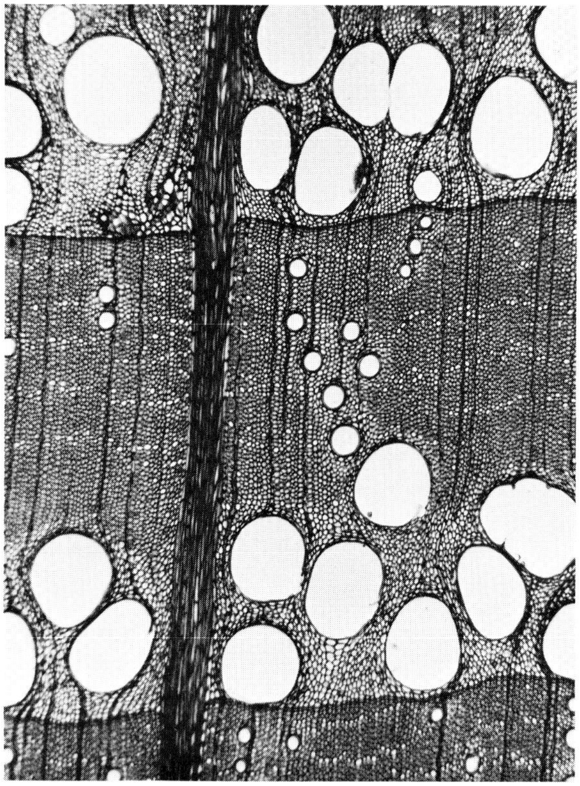

Figure 57 Red oak × 25 Sp. gr. 56

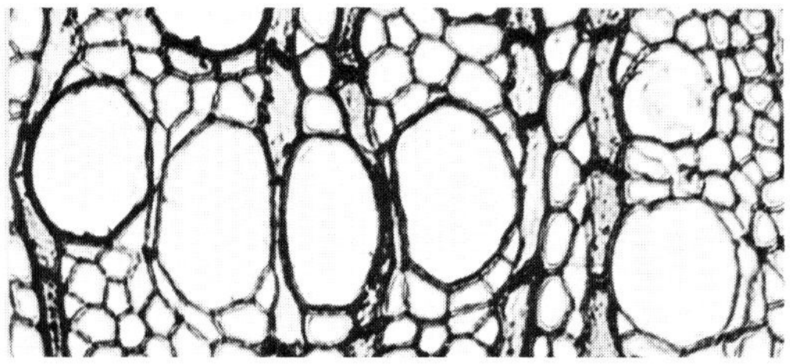

Figure 58 White oak, latewood pores × 150

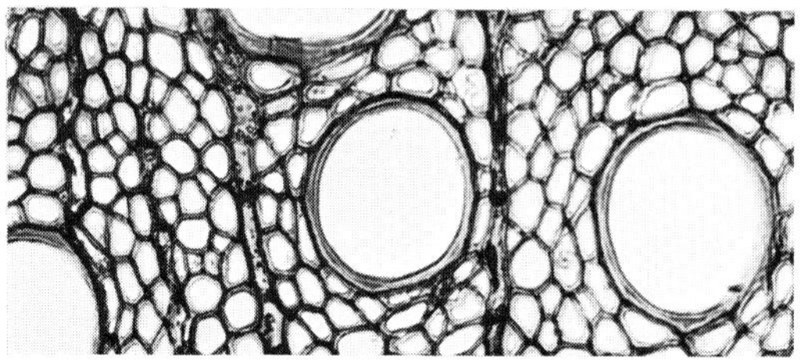

Figure 59 Red oak, latewood pores × 150

White Oak and Red Oak. Each of these represents a large number of separate species, grouped into the "White Oaks", and "Red Oaks". Although the many species (60 to 70 in North America) can be identified by leaves, acorns, bark, etc., it is practically impossible to separate them by wood features, except into the two groups. White oaks have numerous, small, thin-walled, angular latewood pores; red oaks have fewer, usually larger, thick-walled, round latewood pores. In each picture, one complete ring is shown, and parts of two others. The broad wood rays seen in each are found in the oaks, and a few other woods. Very narrow rays are also present.

The wood of oaks is hard and heavy, and especially that of the white oaks is noted for its resistance to decay. The oaks furnish more American timber than any other native broad-leaved trees.

Uses include *railroad ties, veneer, flooring, furniture, ship* and *boat building, barrels* and *other cooperage, firewood,* and many others. Cooperage to hold liquids is made from white oak because its earlywood pores are more or less sealed by tyloses. These pores in red oak are open, and barrels made from it will leak, unless the wood is first treated to close the pores.

Yellow birch is one of the three important northern hardwoods, the other two being sugar maple and beech. In each case, yellow birch and sugar maple represent several other species of birches and maples found in northern, southern (river birch, eastern, and western (Bigleaf maple) United States.

The heartwood of both yellow birch and sugar maple is brown to reddish brown. The picture shows that the pores of yellow birch are the larger, with diameters greater than wood ray width. In sugar maple with smaller pores, the width of the largest rays approximates or approaches pore diameter.

Sugar maple wood has some common uses with that of yellow birch. These include *veneer* (certain trees have beautiful figure patterns and the veneer is used in plywood paneling), *furniture* (where curly and birdseye maple are esteemed), *hardwood distillation* (see beech), *railroad ties* (treated to resist decay), *planing mill products* (especially sugar maple for floors that get hard usage in bowling alleys, and dance halls), *shuttles, spools* and *bobbins; refrigerators, butcher's blocks, woodenware* and *novelties, musical instruments,* and *fuelwood.*

Paper birch wood is used for *toothpicks* and *shoe pegs.* Further uses for sugar maple include *piano frames* and *backs, piano actions,* the *backs* of *violins;* and *athletic goods* (such as billiard cues and bowling pins.)

Figure 60 Yellow birch × 25 Sp. gr. .60

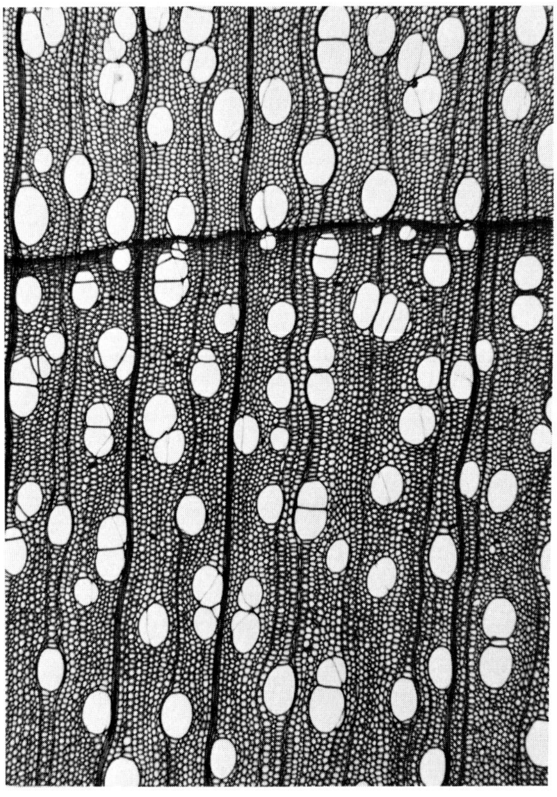

Figure 61 Sugar maple × 25 Sp. gr. .63

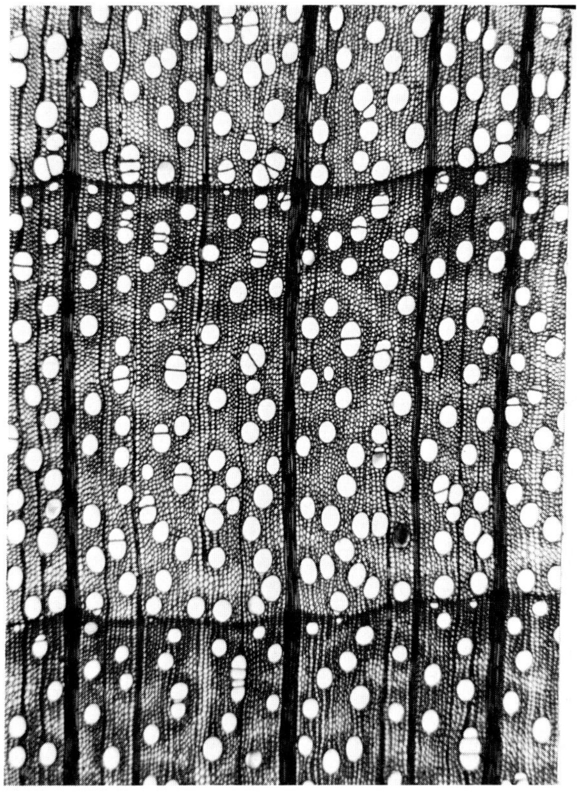

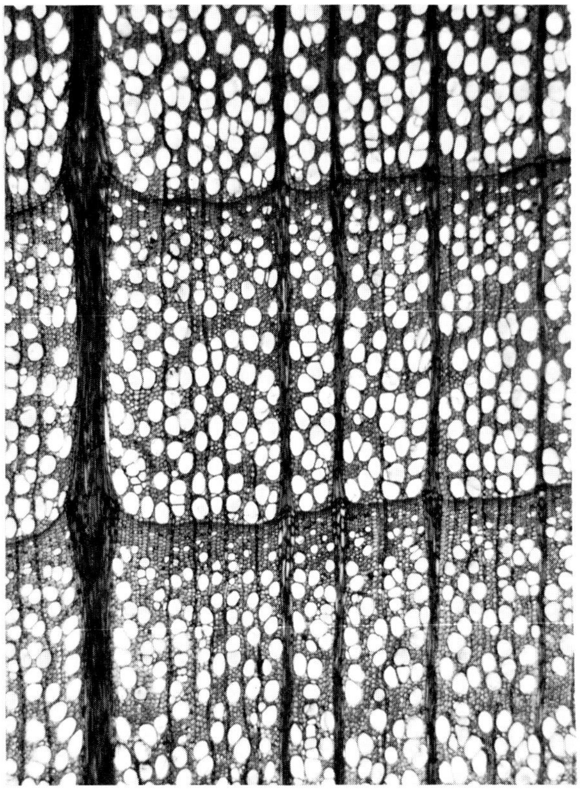

Figure 62 American beech × 25 Sp. gr. .60

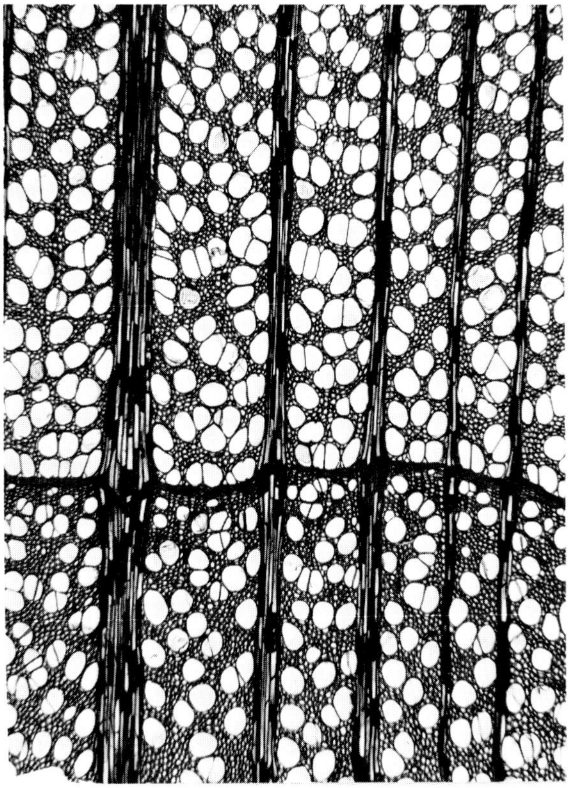

Figure 63 Sycamore × 25 Sp. gr. .47

Both of the above trees range over eastern North America.

American beech heartwood varies in color from almost white to reddish brown. Like the oaks, the rays are of two sizes, either very wide or very narrow. The picture shows 5 broad rays of different sizes. The narrow rays are mostly one to three cells wide and barely visible. Beech wood may have either straight or interlocked grain.

Uses include *planing mill products* (especially flooring), *slack cooperage* (not water tight, mostly for fruit and vegetable barrels), *veneer, fuel wood, furniture, woodenware, handles, railroad ties* when treated with preservatives (beech does not resist decay), and *hardwood distillation* yielding acetic acid, methanol (wood alcohol) and other chemicals. The residue is salable *charcoal*. Maple, and some of the other heavier hardwoods, are also so used.

American Sycamore (Planetree) heartwood is brownish in color, and has interlocked grain which makes it difficult to split. The largest wood rays are about 14 cells wide, and ray width, as shown, grades downward to rays of one or two cells in width. When a log is quarter sawn, the conspicuous crowded rays display a beautiful surface pattern.

Uses include *veneer* for vegetable and fruit baskets, *boxes* and *crates, slack cooperage* (flour and sugar barrels) since the wood will not stain the contents or transmit odor or taste, *millwork* (interior paneling and trim), and *furniture*.

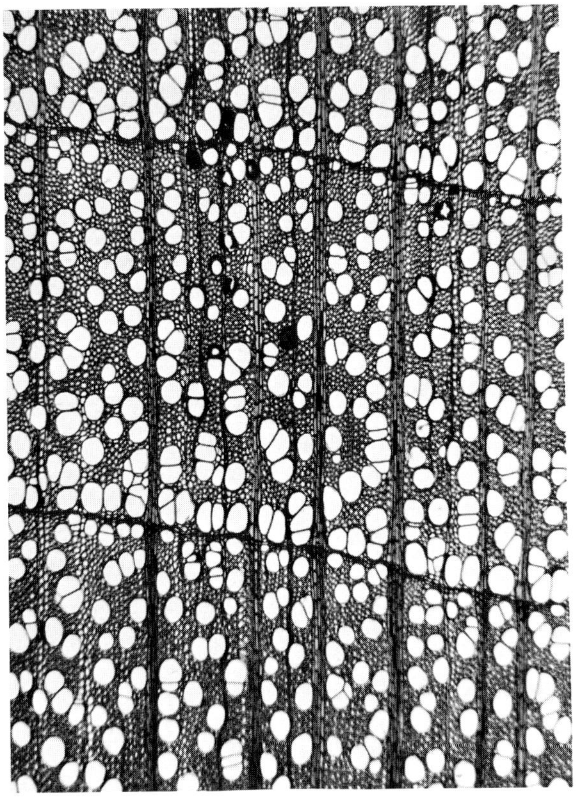

Figure 64 Black cherry × 25 Sp. gr. .48

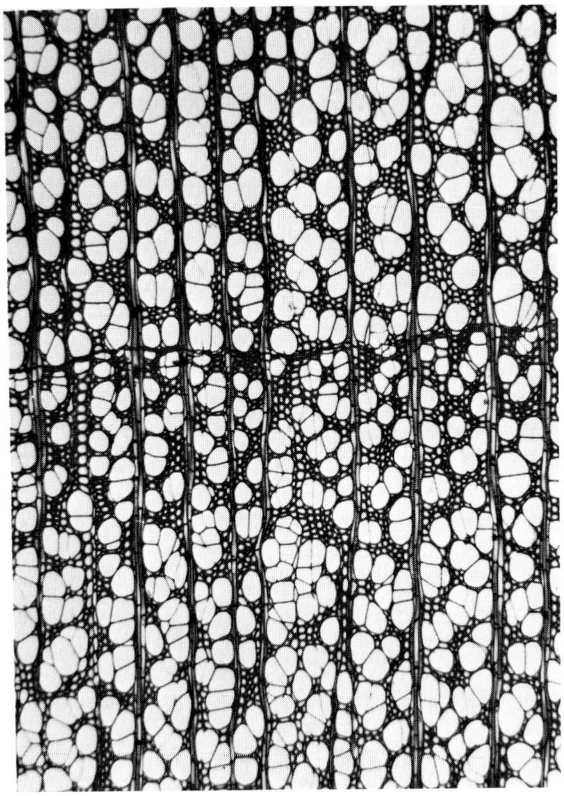

Figure 65 Yellow-poplar × 25 Sp. gr. .40

These two trees range over the eastern United States. Black cherry also occurs west of the Mississippi into Texas, Arizona, and south through Mexico.

Black Cherry like black walnut is a superb cabinet wood because it has low shrinkage, does not check or warp, and the finished heartwood has a beautiful reddish brown color and luster.

Both colonial and modern *furniture* made from black cherry wood are highly prized. The best grades of cherry are always in short supply. Other uses include *printer's blocks* to which are attached electrotypes, *interior trim* (in cars, houses and boats), *piano actions, scientific instruments, paneling, handles,* and *woodenware.*

Yellow-poplar (tulip-poplar) is not a poplar at all but belongs in the Magnolia family. More confusing still, it is often called simply "poplar". The heartwood is quite variable in color, yellowish to tan, or greenish brown. If desired, it may be darkened by soaking in water.

The uses include *veneer* to make *plywood* for *furniture* or *paneling*. The wood takes glue well and is used for the interior or core piece of a plywood sheet. Beautifully patterned wood from burls is used on the outside of plywood panels. Some other uses are for *woodenware, interior trim, turnery,* and *cabinet work.*

Both of these trees are essentially southern species extending, however, northward along the Atlantic coastal plain, and in the Mississippi Valley.

Sweetgum (redgum) sapwood is white with a pinkish cast. It is sold in this country as "sap-gum", and in Britain as "hazel pine". The reddish brown heartwood is often streaked with a pigment figure, and then it is called "figured redgum". This resembles English or Circassian walnut, and is sold abroad as "satin walnut" some of which, made into furniture, is sold back to us in the United States!

The uses include *veneer* (in which it is second only to Douglas-fir) for *plywood* and *panels; fruit boxes* and *crates, mine props, pulpwood, railroad ties, furniture, planing-mill products, slack cooperage* and others.

Holly is the only North American hardwood to have nearly white heartwood. Actually it is ivory-white often with a bluish or greenish tinge. The largest wood rays are distinct, as well as the pores that occur in radial "strings".

The uses include *inlay work* (furniture), *handles, brush backs, carvings, measuring scales,* and *rulers, bread boards, engravings* and *scroll work*. The wood may be stained black to imitate ebony and is then used for piano keys.

Figure 66 Sweet gum × 25 Sp. gr. .48

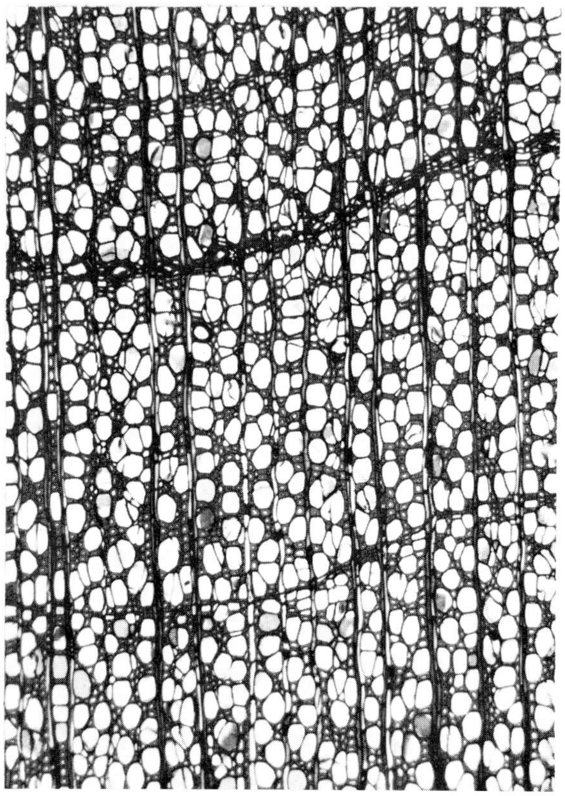

Figure 67 Holly × 25 Sp. gr. .50

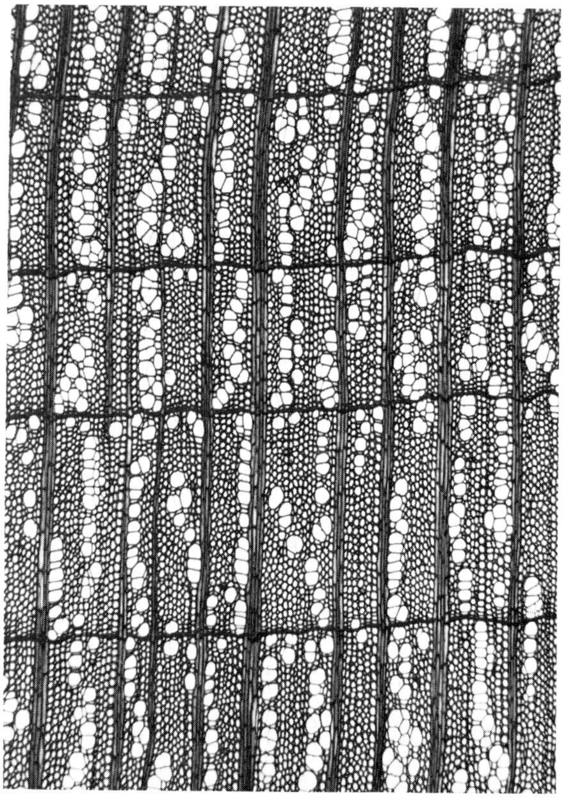

Basswood includes two or more species found variously through southern Canada, central, eastern, and southern United States. The wood has a slight but characteristic odor from a freshly cut surface, especially when wet. The heartwood is pale brown. Although light in weight it has a certain toughness which is utilized in the famous basswood *canoes, honey boxes* for comb honey (a single slat is dovetailed at both ends, scored with a V-notch for the 3 other corners, and bent without breaking the very thin portion left there), and *trunks,* and *valises* (mostly as plywood). Other important uses are many including *veneer* for *plywood* used as drawer panels and other hidden parts in *furniture,* plywood *core stock* to which is glued veneer from expensive cabinet woods, *excelsior, millwork,* especially sash and doors, *venetian blind slats, piano keys,* and *boxes* and *crates.*

Persimmon is essentially a southern tree, but extends northward in the Mississippi and Ohio Valleys and across Pennsylvania to New Jersey. The creamy white sapwood turns to a yellowish brown upon exposure, and since the brown to black heartwood is very small in diameter, most of the commercial stock is sapwood.

The heavy, hard, tough wood stays smooth under friction, and is therefore used for *shuttles, spools,* and *bobbins* in the weaving industry, *handles, golf club heads,* and *shoe lasts* (now mostly replaced by hard maple, a cheaper wood).

Figure 68 Basswood × 25 Sp. gr. .35

Figure 69 Persimmon × 25 Sp. gr. .64

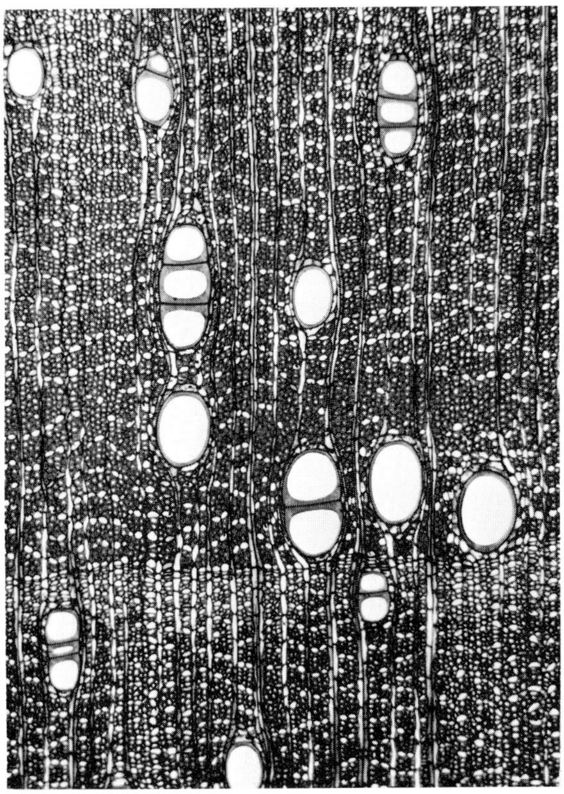

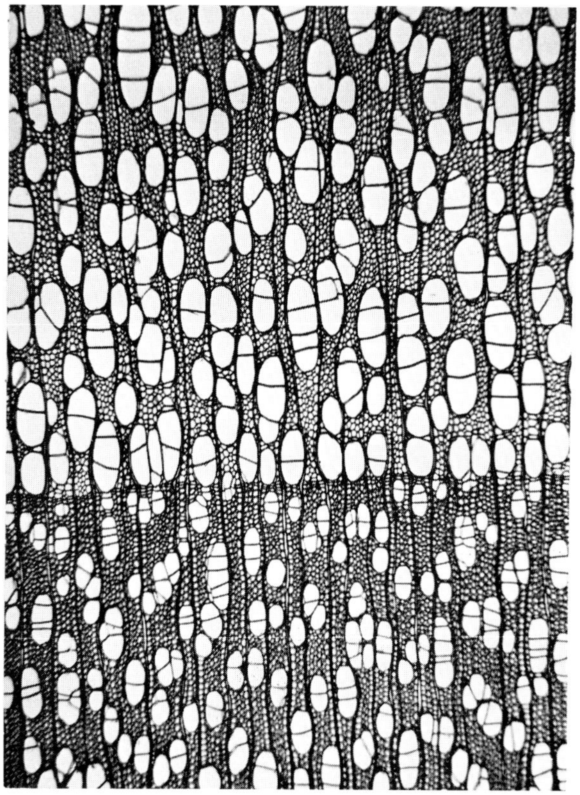

Figure 70 Cottonwood × 25 Sp. gr. .35

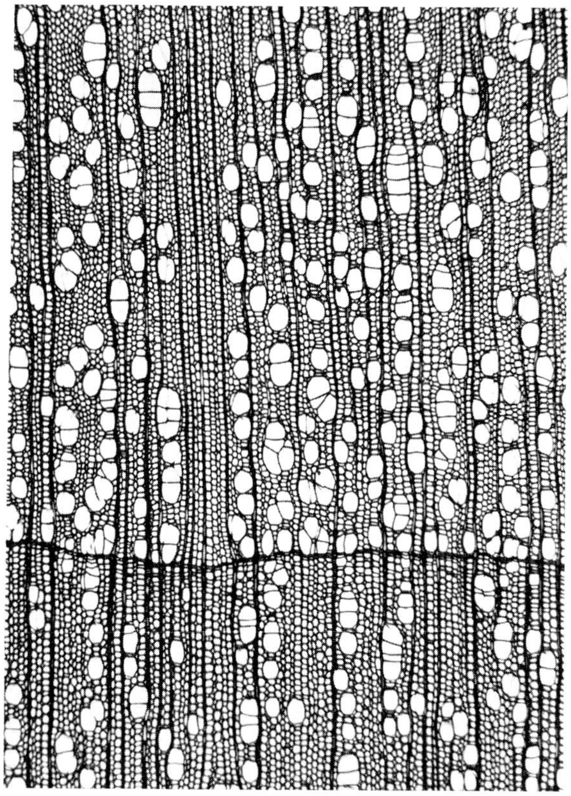

Figure 71 Red alder × 25 Sp. gr. .40

Cottonwood is derived from several species found over considerable areas of North America, north, south, east, and west. The heartwood is light grayish brown. Since these are fast growing trees, the growth rings may be up to an inch wide.

Uses include *veneer* mostly for the core, or cross-banding sheets in *plywood* for *furniture, wood pulp* for high grade paper used in magazines and books, *excelsior* (with the aspens, these are the principal woods for this purpose because of light color and weight, and easy shredding properties). The lumber is excellent for *boxes* and *crates* (it nails well without splitting, and its light color recommends it for stenciling. Other items include *tubs* and *pails* for lard and butter.

Red alder is the principal hardwood of the Pacific Northwest not only on account of its good qualities, but also because western forests are mostly coniferous (softwoods), and the supply of hardwoods is small. Red alder wood is whitish when first cut. Upon exposure it darkens to a light reddish brown. The heartwood is not distinct. The picture shows (on the left) that sometimes several wood rays grow closely together forming an "aggregate ray." This feature may be more conspicuous than shown, but occurs irregularly.

The wood works well, takes glue, holds nails and screws, and finishes display a pleasing grain. These features recommend it for *chair* manufacture and other *furniture*. Further uses include *plywood* (core stock and crossbands), *wood pulp* (important), *millwork* (sash and doors), *woodenware handles*, and *charcoal*.

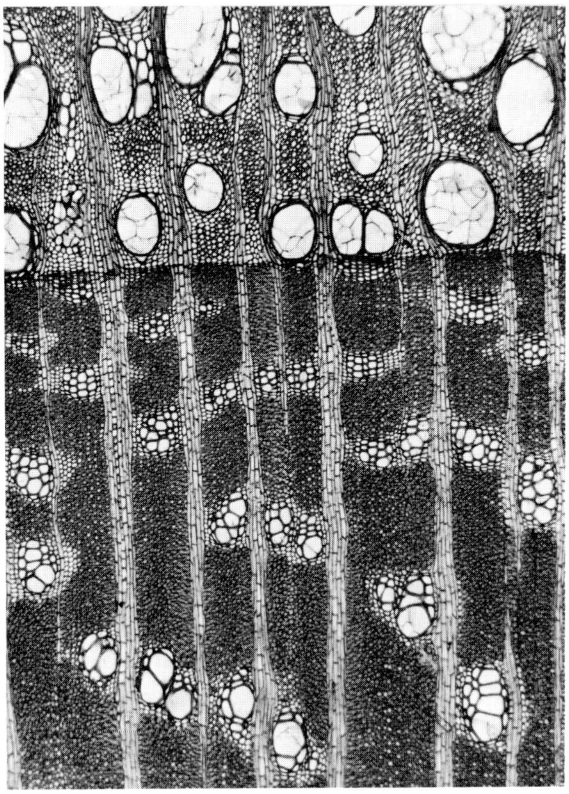

Figure 72 Fast Growth Black locust × 25 Sp. gr. .66

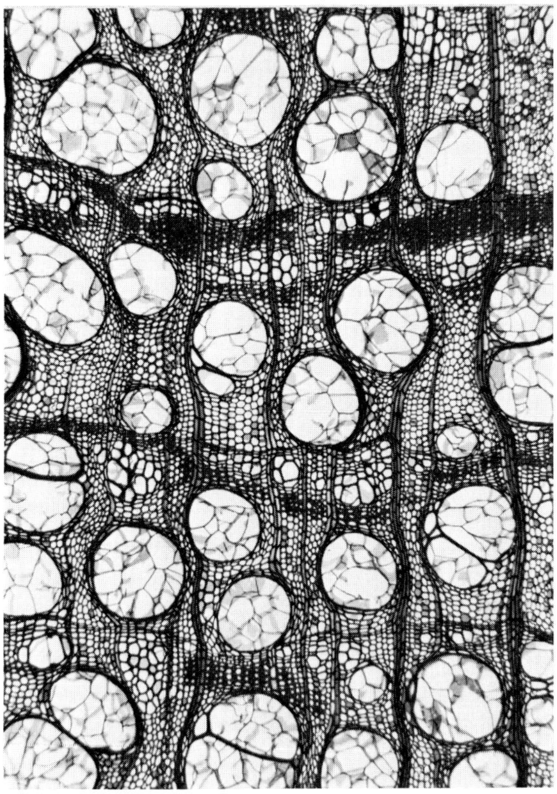

Figure 73 Slow Growth

Black locust. Heartwood: greenish yellow to golden brown, earlywood vessels completely filled with tyloses, latewood vessels in nestlike groups arranged in interrupted concentric bands. Uses: *Insulator pins* have been the chief manufactured product. The wood is especially valuable for this use because it shrinks and swells very little with changes in moisture, and also resists wood-decaying fungi. Black locust wood is also used for *mine timbers, fence posts, railroad ties,* and in construction where hardness, strength, and durability are most important.

The two sections above were chosen to show how rate of growth affects structure. At first glance it might seem that they are not even of the same wood. The one on the left shows latewood (lower two thirds) of one season, and above it the earlywood of the next. Notice the great masses of *fibers* that give strength to the wood. They are so small and have such thick walls that individual ones can scarcely be seen at this magnification. The section on the right shows at least three rings and perhaps more, but the pattern is difficult to interpret. There is very little latewood showing fibers. This section illustrates that in some ring-porous woods, earlywood vessels may make up most of the season's growth; there is very little latewood. Which piece of wood do you think is the heaviest, hardest, and strongest?

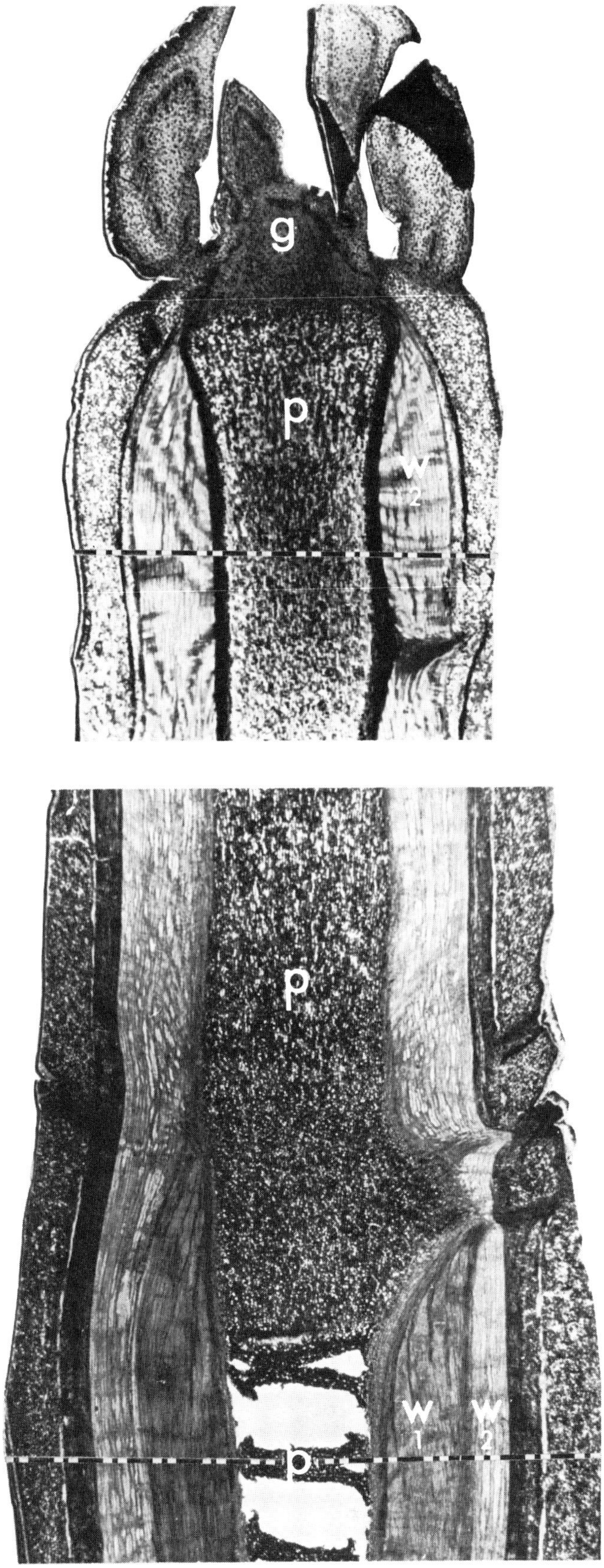

Figure 74 Twig of butternut; radial, and cross sections × 6. g, growing point. p, pith. W1, first wood formed at this height. W2, next year's wood. The dashed lines are the planes of section.

Chapter IV

How Trees Grow Wood and Bark

Twig Growth in Length

In telling the story of tree rings, we have not yet seen how these layers appear in a vertical section through the center of a tree. The sections of a butternut twig two years old were made in autumn after growth had stopped for the season. Several inches of twig have been cut out between the upper and lower vertical sections. If this were not done, the twig picture would be two or three times the height of the page.

At the top, the *growing point,* "g", is enclosed by several thick bud scales. These protect it from drying out during the long winter. Notice that all the tissues below, including pith and wood seem to "flow" downward from this end zone. During the growing season, the cells here continuously and rapidly divide, and as they elongate, the tip of the twig grows upward.

Now look at the lower half of the twig, and discover that there are two layers of wood, W1 next to the pith, and W2 extending past W1 nearly to the tip of the twig. Each year a new layer of wood is added to that or those already formed, while during the same season, this new layer extends the twig's growth further upward into space. Among other layers that develop from the growing point is the *cambium* between the wood and inner bark. Cell divisions in this layer produce the wood that you see in the pictures.

A year before the twig was cut from the tree, the growing point was of course just above where the top of W1 meets the pith. Therefore W1 is a year older than W2. If the twig had not been cut, the growing point during the next growing season would have advanced, lengthening the twig, and the cambium would have produced layer number three (W3) telescoping W2.

The cross sections are not from the same twig sectioned lengthwise. How could this be done? Rather, they are from a similar twig to show general relationships between the twig sections in both vertical and cross section. It is especially noticeable that the pith in the lower cross section happens to be larger than that in the other twig. The dot and dash lines show the locations of the cross sections.

Walnuts (including butternut) have an unusual type of pith. In autumn, the previously solid pith divides crosswise forming partitions. This had not yet taken place when the twig was collected, but shows clearly in the last year's growth.

Finally notice that in the lower cross section W1 is next to the pith while in the section above, the wood here is W2. Just remember that in point of time when W1 (vertical) finished growing, W2 did not yet exist.

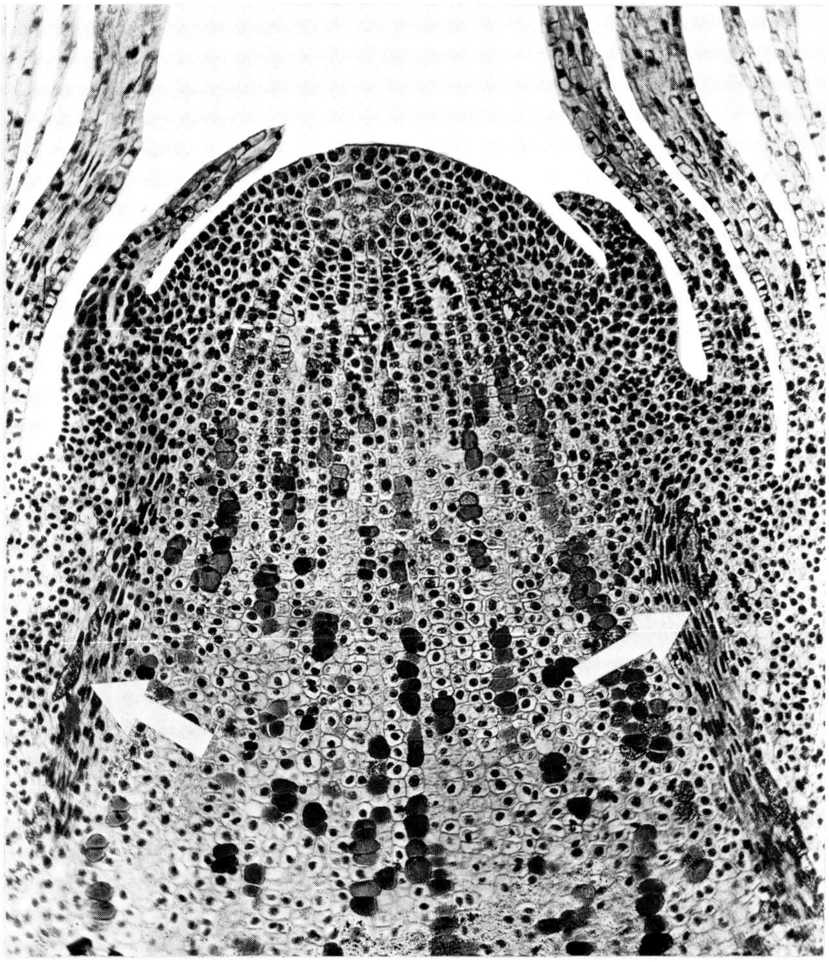

Figure 75 Growing Point of White Pine × 100. Arrows show Pro-cambium.

All of the many and often beautifully sculptured cell types in a mature woody stem originate indirectly or directly as cells, of about the same size and shape, in the growing points at the ends of every live twig and root. Cell division occurs at and just below the growing tip. Below this region is the zone of elongation where certain groups of cells grow exceedingly in length. Their final length may be 100 times greater than it was when elongation began. In the great central mass below the growing point, extending to the bottom of the picture, little change takes place in cell shape because this region is to be the pith at the center of the twig, and pith cells are more or less the same size and shape. Tree growth in length occurs only at twig and root tips.

On each side of the pith is a zone of elongating cells called the *procambium*. If we could follow the center cells down the twig we should find that they soon become the cambium. Were it not for the cambium there would be no growth in diameter of trees and shrubs.

Few people have the slightest idea of how trees grow both in height and diameter. If you drive a nail into a tree, at six feet from the ground, and the tree grows two feet in height each year for five years, how far will the nail be above the ground at the end of the five years? Of course you now know that it will be right where you drove it, but this question confuses many people who feel that the nail must have moved up as the tree grew.

Growth in Diameter

The ancients knew that a tree was made up of bark on the outside, and wood inside. They could see that on large trees, there was an outer layer of dry dead bark, and an inner one somewhat greenish and moist, the inner bark. They found that in certain trees the stripped inner bark was very tough and useful for binding, and upon soaking, yielded pliable bundles of fibers for rope making. From a verb "to bind" came the word *bast* or *bass*. Basswood was so named because its "bast" or inner bark produces excellent fibers. The Chippewa, Iroquois, and other woodland Indians made superior hand twisted rope and string from basswood and elm bark cooked in a solution of wood ashes, or left to soak for a month or more in a pond.

Presently, botanists refer to the dead outer bark as such (or rhytidome), and use the term *phloem* (Greek word Phloos, meaning bark) for the inner living bark tissues.

Ancient man also knew that, in the spring, bark could be peeled easily from a tree, but as summer advanced, this became more difficult. By autumn, and during the winter, to peel the bark was difficult or impossible. In spring, there seemed to be a sort of thin, moist or semi-fluid layer between the inner bark and the last formed wood of the previous season. Long before the first microscopes revealed the cell structure of wood, bark, and finally cambium, this layer was recognized and so called (from the Latin "cambiare", to exchange). It was thought that the cambium was composed of a "humor" (Latin, meaning fluid) which in some mysterious way produced wood inwardly, and bark outwardly. Even after the first microscopes revealed that plants are made up of innumerable cells, the nature of the cambium may for a time have eluded early botanists. Bark, cambium, and wood are very difficult to section together. Between the relatively hard bark and wood lies the delicate tissue of the cambium. The best cutting knife available was the old straight razor and, with this, remarkably thin sections were cut by hand. But the cambium was prone to tear, and little detail could be seen in it. However, with more practice, especially on samples taken in the winter condition, the cambium was finally revealed as one to several layers of living cells separating wood and inner bark (phloem). There was complete radial cellular transition from one to the other, with no "humor" at all in between (p. 58).

Before looking at the cambium in greater detail with a microscope, it is important to visualize it in all three dimensions at once. Actually the cambium is a sheath of billions of cells between the inner bark (*phloem*), and wood (*xylem*), which covers the *entire body of the tree* except at the very twig tips (p. 54) and root tips. Between the phloem and the xylem is the cambium. Without it there would be no phloem or xylem. We are so used to seeing and studying cross sections of twigs, branches, and stumps where these three layers appear as patterns of concentric rings, that we hardly ever wonder how these growth layers would look in a tree trunk cut lengthwise through the center from top to bottom. These layers have already been shown in a two year old twig (p. 54). One may think of each year's new layer of wood as a narrow hollow cone which telescopes over the growth of the previous year.

The Cambium

For the first time in this book, you have a picture showing the phloem, cambium, and developing xylem, during the growing season.* At the top are food conducting cells, the *sieve tubes*. Just below is a tangential band of tannin filled cells, and below them are more sieve tubes and scattered tannin cells. The latter mark the lower boundary of mature phloem. Notice that the rays generated by the cambium are a part of the phloem as well as the xylem.

Below this is the cambial zone of dividing cells. Further below, these become larger radially and finally develop the relatively thick secondary walls of the mature tracheids. Polarized light indicates the cambial zone with the maturing phloem above, the maturing xylem below. By the time the tracheids have become thick-walled, the

**See Figures 76 and 77*

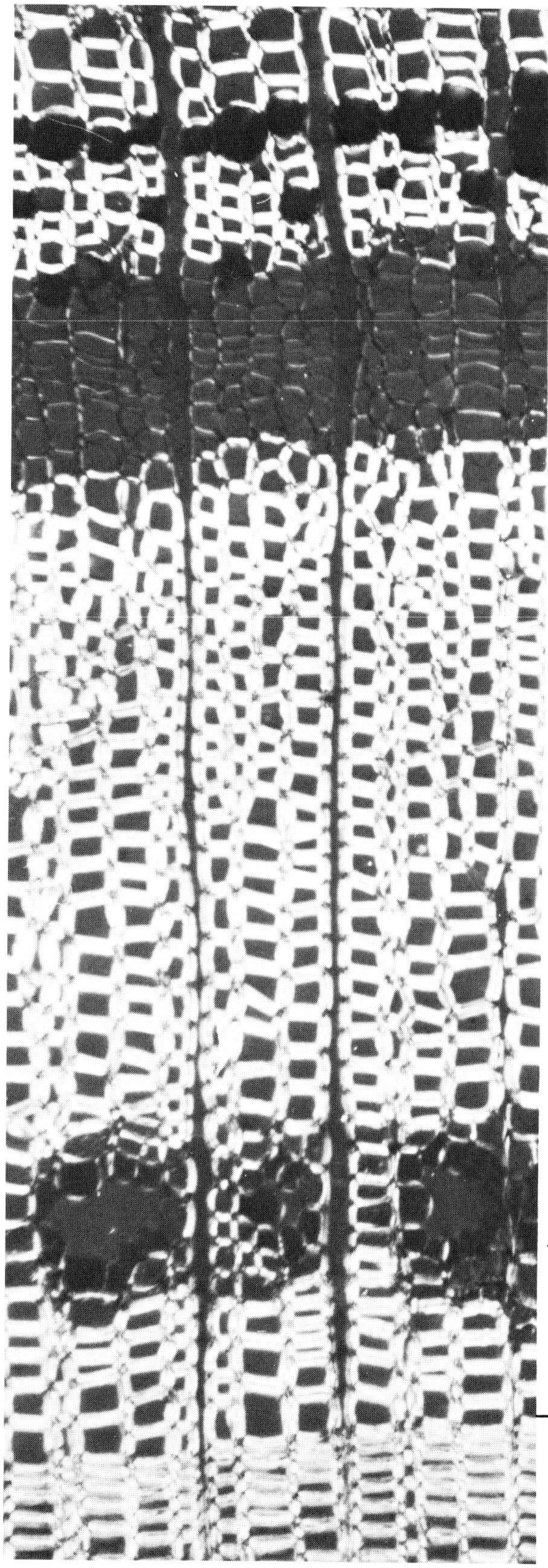

Figure 76 With polarized light

Figure 77 With bright field illumination

protoplasm within each of them dies, and the tracheids remain as hollow water-conducting tubes. However, the wood ray cells remain alive and serve to store organic food materials until heartwood formation occurs when they, too, die. Also the lining cells of the resin canals are alive and exude resin during the life of the sapwood.

At this point, notice that the phloem, cambium, and xylem are all composed of radial rows, or files of cells. Except for the interruption of the resin canals, you can start at the top of the picture and, with some exceptions, follow a single file of cells all the way from top to bottom. Each of these files came directly, or more often indirectly, from a single cambial cell called an *initial.*

On page 57 it was stated that the "cambium was finally revealed as one to several layers of living cells. . . ." If you look carefully at the *cambial zone,* it is apparent that the cells look so similar that it is not possible to distinguish a single tangential row of them different in appearance from the others—the *true cambium* Because of this difficulty, some researchers call the entire band of dividing cells the "cambium", while others insist that the term be applied to a, single tangential row of *cambial initials,* the true cambium.

Considering the cambium to be a single tangential row of initials capable of repeated divisions year after year for the life of the tree, let us see what goes on during the growing season, in a radial file of cells. If we understand this, we can assume that the same process occurs in the other files as well.

The cambial initial divides, and a new tangential wall appears across its center (a few of these new walls can be seen in the picture). Now there are two cells instead of one. One of the cells remains the cambial initial. The other one may mature directly into a tracheid, or into a sieve tube. Which it will be depends upon its *position* in relation to the other cell which remains a cambial initial. In this two cell situation, if the outer one (toward the bark) remains cambium, then the inner one becomes a tracheid. Similarly, if the inner one remains cambium, then the outer one becomes a sieve tube. However, it is usually not as simple as this. The cells from the division of cambial initials are called "mother cells" because they themselves can again divide from one to several times. This speeds up greatly the production of new xylem and phloem.

Red spruce cross section showing phloem (ph), active cambial zone (c), new xylem (nz), resin canals (rc), and last year's latewood (lw). × 230. Micro slide by L. W. Rees.

Fortunately for us, there is much more wood in a tree than there is bark. This is because the cambial initials produce more xylem mother cells than they do phloem mother cells. The picture of the three layers in the winter condition may suggest this situation (See Fig. 78). The cambial zone here is four cells wide (radially). This is best seen in the file at the left. The cambial initial is the top one of the four, next to an immature phloem cell filled with protoplasm, above. The three cells below the initial are xylem mother cells that divide to form tracheids, when spring arrives. This picture like most others of wood structure in this book were made from sections cut one one-thousandth of an inch thick, or less. Live cells such as those of the cambium are very delicate. During preparation, the protoplasm which filled the cells may shrink away from the walls, and these may twist or buckle out of shape. In the mounted section, the cell contents may have disappeared entirely. These things can be seen in the picture.

That bark is most easily peeled from a tree in spring has already been mentioned. One might suppose that "peelability" coincides with the beginning of cambial cell division, but it has been found that the ability to peel may precede cambial initial division, by as much as a month. Before cell division actually begins, the radial walls in the cambial zone become much thinner than in the winter state, and grow radially. Such walls are easily ruptured and peeling can begin. Peelability moves down the tree from the swelling buds. It is presumed that auxins (growth regulators) in the developing, opening buds and new leaves move downward and trigger the resumption of growth in the cambium but there is much yet to learn.

If you consider the circumference of a twig compared to that of a mature tree, you may wonder what happens to the ring of cambial initials as the circumference increases year after year. For a number of years the initials increase in size both tangentially and also in length. This finally ceases and, thereafter, the initials maintain a more or less constant size. During the life of a tree, as its branches, trunk and roots keep getting larger, something else happens in the cambium to keep up with the ever growing circumference. Occasionally, in addition to the usual tangential cell divisions, there is a *radial* one. This results in a new cambial initial that enlarges the tree's circumference by one cell. Repeated "as needed" such radial divisions from year to year keep adding new cells to the ring of cambial initials. On page 58, notice that above the center resin canal, the xylem between the two rays consists of three radial files of tracheids. Follow these up, and about half way to the cambial zone, a new narrow tracheid intrudes. Now there are four files instead of three. From here on up to the cambial zone, the new file of tracheids broadens to about normal tangential size. Notice too the file of tracheids at the left side of the picture. Going upward, the single row ends in an unusually large tracheid followed by small ones. In fact this whole section of tissues seems to have gone through a period of adjustment in which the final production of unusually large tracheids was suddenly followed by smaller ones as several cambial intials divided radially. The same adjustment took place in the phloem where the largest oldest sieve tubes at the top were followed by smaller ones below.

Wood and bark structures in the conifers are relatively simple compared to those in the broad-leaved trees. At this point you might look again at the hardwood pictures on pages 44 to 53. The vessels seen in cross section as pores are

Figure 78 Tamarack phloem, cambial zone, and xylem in the resting winter condition × 500

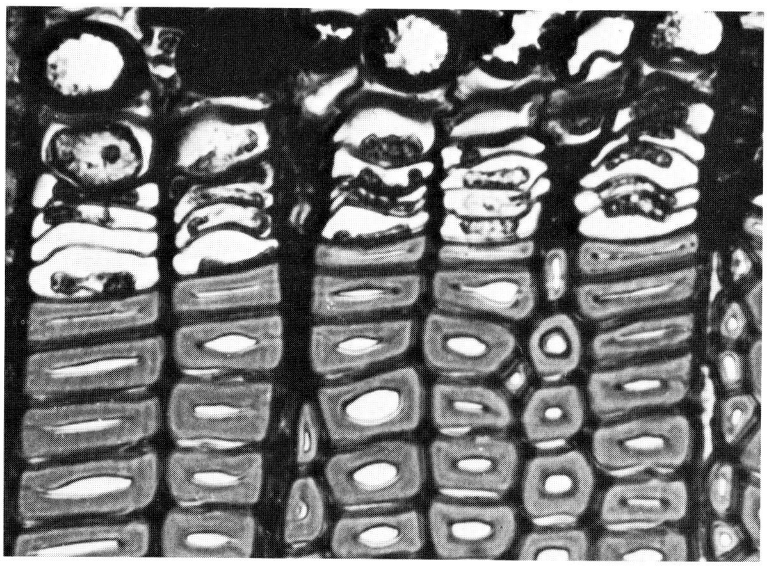

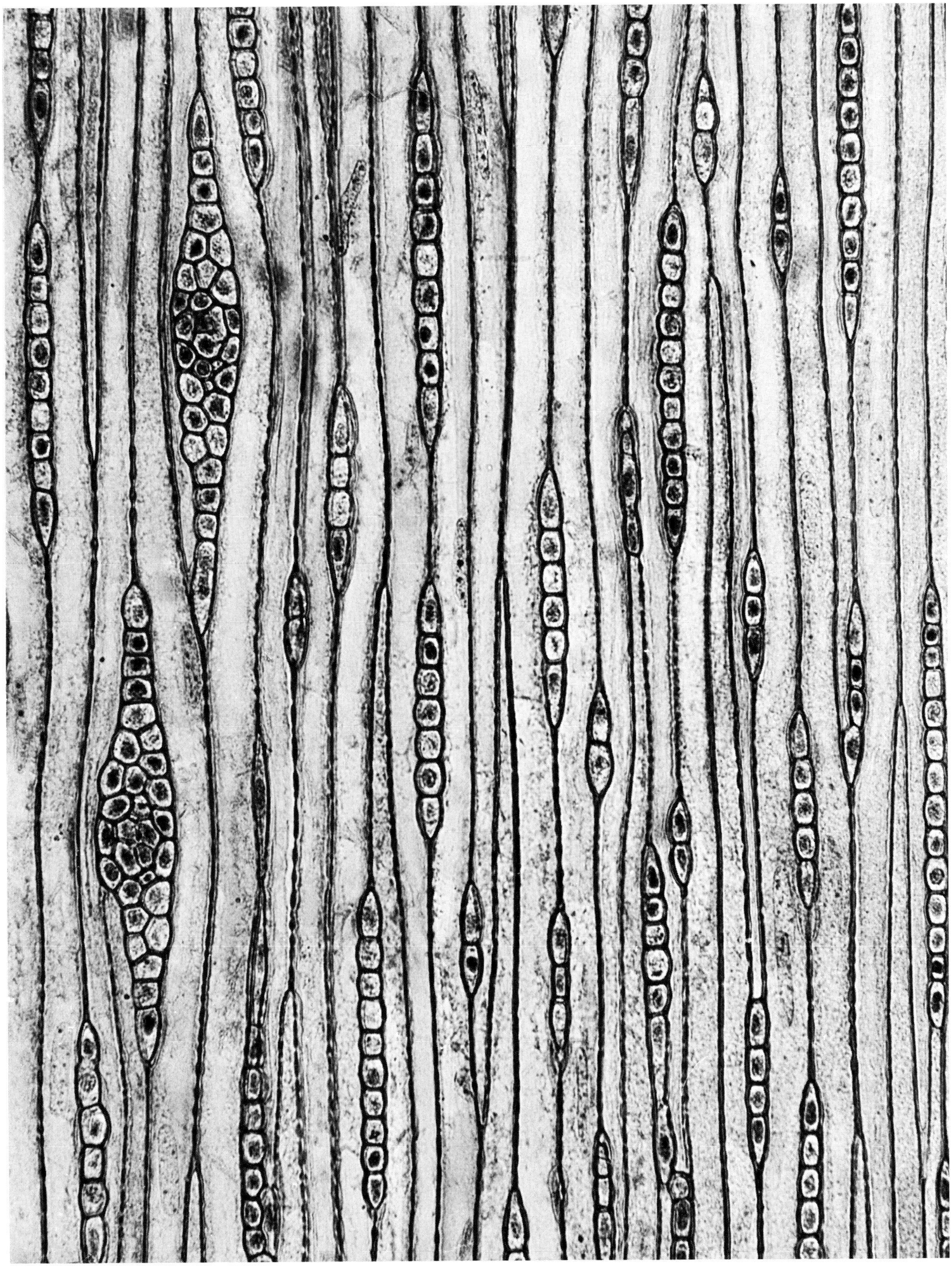

Figure 79 White pine, tangential section through cambial zone × 250

enormous compared to the tracheids or fiber tracheids surrounding them, and yet all three of these cell types were produced from cambial initials of about the *same size.* The tremendous enlargement of a cambial initial to produce a vessel segment is best seen in the earlywood vessels of ring porous woods where the diameter of a vessel may be 20 to 50 times that of the initial from which it grew. How can such enlargement take place against the pressure developed by the surrounding tissues? We know little about this, but it is clear that as an initial begins to enlarge, some others "drop out" or fail to form, giving it more room. Even so, there is often great distortion of the surrounding tissues. This is seen especially on pages 44 and 53, where the wood rays are pushed sideways by the expanding vessels.

Before leaving the cambium, perhaps the most remarkable layer of dividing, growing cells in the plant world, we should see how it looks tangentially (as though one were looking from the outside directly toward the pith).* Sections cut one one-thousandth of an inch thick may include a layer of initials and mother cells not more than one to three cells thick. To position a block in the cutting machine (microtome) so that one gets perfect alignment of a section through so thin a layer as the cambial zone takes great skill, patience and luck. Remember too that the circumference is more or less curved depending upon the size of the tree from which the block sectioned was taken. This adds to the difficulty of making a good section.

Tangentially, you can see that the cambial cells are of two shapes, the very long ones with tapered ends that mature into tracheids or sieve tubes (white pine), and the bead-shaped ones in scattered strings that form the rays. At the left are two somewhat spindle-shaped groups each of which develops a ray with an included horizontal resin canal. Some of the long cambial cells appear as narrow channels open at both the top and the bottom of the picture. At this magnification, their length would be about two feet and therefore the overlap zone does not show.

Although the picture does not show it, the walls between adjacent cells are three-layered. Each cell has its own *primary wall,* and between this and the wall of the adjacent cell is the intercellular substance or middle lamella.

Near the center of the picture is a ray on both sides of which are long cells with narrow tapering, overlapping ends. When mature tracheid lengths in softwoods are compared with their cambial initial lengths, it is found that the tracheids in maturing become 25-30% *longer* than the initials. This has been a source of puzzlement for many years. If you look at the tapered ends and assume that the intercellular substance is soft and yielding, it may be that during the growing process, these ends push lengthwise, and grow between the cells above or below them. There are objections to this idea, but other theories are also open to question. In certain of the hardwoods, long thick-walled fibers may be four or five times longer than their cambial initials.

Cell division in the long cambial cells is one of the most remarkable achievements in the plant world. This is because of the enormous relative length of these cells some one hundred times longer than they are across. The beautiful complexity of "ordinary" cell division (see a textbook of botany) is here "stretched to the limit" in these long cells. You can see the elongate nuclei as shadowy bodies in some of the cells. The nucleus is about half way from one end to the other. A nucleus divides lengthwise resulting in two nuclei between which a new wall forms. Then the wall begins to extend lengthwise from the middle toward both ends of the long cambial cell. At the peak of growth in spring, this new wall may be built in as short a time as 24 hours.

Many lifetimes of research have been spent in discovering the secrets of the cambium, and in this short sketch only a few of the important aspects have been described. For those who would go further, the following reference is recommended (8).

* *Figure 79*

Figure 80 Sugar Maple. 1. outer bark, 2. inner bark (phloem), 3. sapwood

The Cork Cambium

Although the way in which inner bark changes to outer bark is not actually a part of the wood story, it seems near enough to be mentioned. The inner bark of a freshly cut tree is light in color, but when exposed to the air for some time it may darken as shown above.

In the outer layers of the inner bark (phloem), from time to time, groups of living cells become capable of cell division. Such a layer is called the *cork cambium*. It produces cork cells outside, and thin-walled cells inside. The three layers together are the *periderm*. Old periderms can be seen in the outer bark. These developed in irregular arcs or lines which are not continuous around the tree. Everything outside the last formed periderm is dead and serves only to protect the trunk from fire (except birches) and other injury. The thickness of the outer bark on different kinds of trees varies exceedingly depending upon weathering, cracking, and other causes. On the giant sequoia (bigtree) of the Sierra Nevada, the bark may be from one to two feet thick.

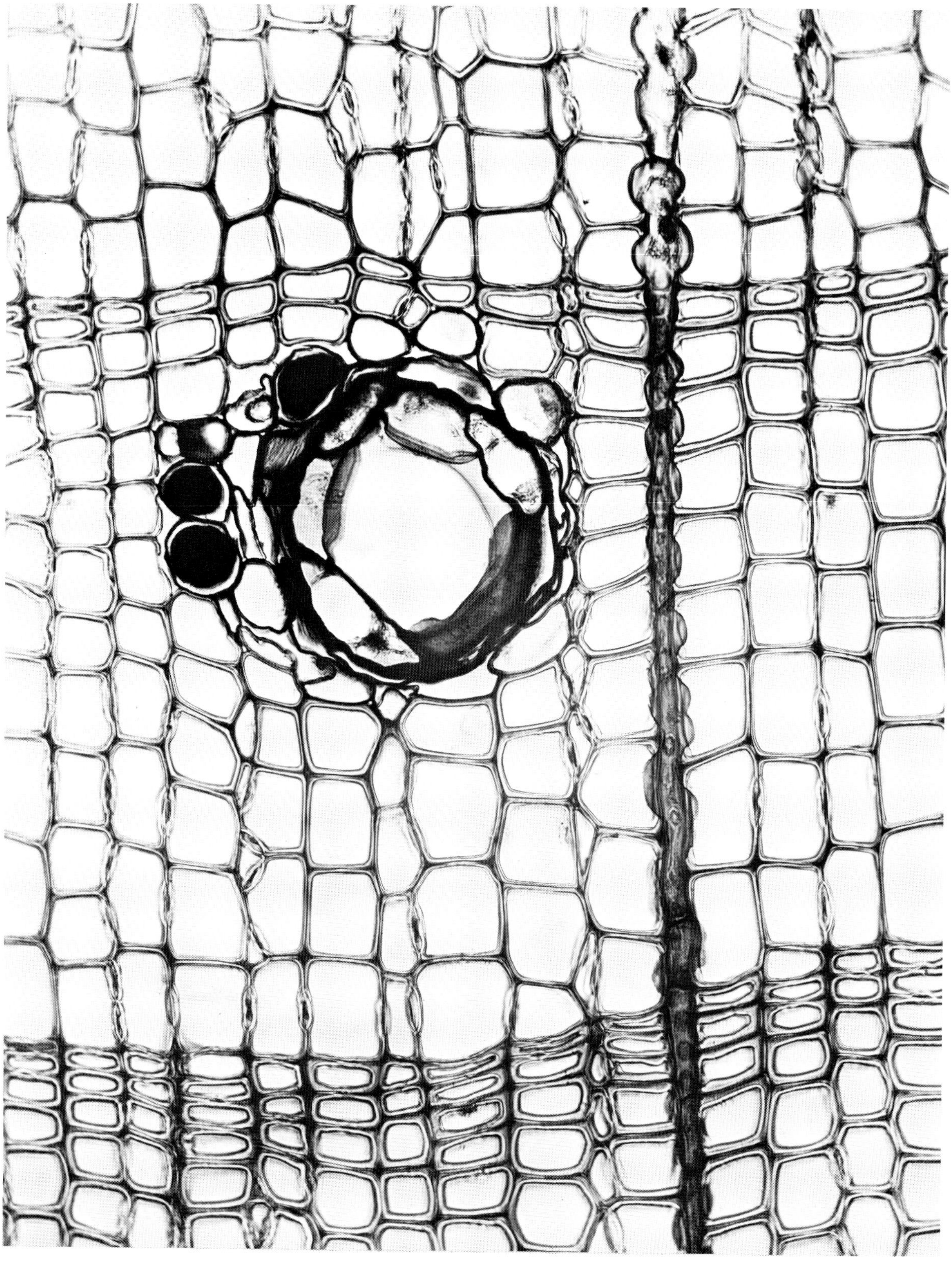

Figure 81 White Pine cross section × 300 showing one complete growth ring, a wood ray, and one of the large resin canals that characterize the pines.

Chapter V

Wood As A Material

An Overlook

When plants first crawled out of the ocean, more than 400 million years ago, certain of them began to develop upright stems featured by a remarkable tissue that we call wood. We have already seen how wood looks under the microscope, and this chapter gives an overlook of the unique properties of wood. Its versatility surpasses that of any other material in the world. In fact any man-made product having even a few of the many characteristics of wood is hard to imagine and would be considered a great scientific achievement. Perhaps the reason we do not appreciate wood is that it is so common, but often common or ever-present things are indispensable; consider the sun's heat and light, air and water. To these should be added wood because trees continue to grow it, year after year, crop after crop. Substances mined from the earth will finally come to an end, but man will always have trees and their wood.

Perhaps we should first dispose of wood's so-called "bad" features. (1) Wood shrinks across the grain when dried from the green condition, but the lengthwise shrinkage is negligible, and this is often the most important dimension where wood is used. To minimize shrinkage problems, green wood should be dried to about the moisture content that it will experience in use *before* it is manufactured into furniture, interior finish, flooring or other items. There are treatments that make wood much more stable than normal for use as window frames and other things where the moisture content may vary considerably. Another way to minimize shrinkage across the grain is to use manufactured plywood in which thin layers cross each other at right angles. This results in a glued sheet of wood with almost no shrinkage across the face in any direction. Finally, it should be noted that "nature" has no use for dry wood. In the living tree the cell walls are saturated with water even though there may be no free water in the cell cavities. Under these conditions there is no swelling or shrinking of the wood. (More details on this subject appear on page 68.)

(2) Wood decays, but only if it is moist so that wood destroying fungi can live in it. Keep wood dry and it will practically last "forever". Timbers 2,700 years old have been found in the tomb of King Gordius near Ankara, Turkey. Ancient Egyptian tombs have also contained wooden articles of great age. These are, of course, especially dry places. Designs for wooden structures should always provide for drainage, and not allow water to seep into cracks; also untreated wood must not touch the ground; in fact it should be at least 8 inches away from it. Completely saturated wood, with the cell cavities filled with water, also does not decay because there is no oxygen which fungi need to grow. Over thousands of years it does change chemically and loses strength.

(3) Wood burns. Thin shavings catch fire at once, but the larger the piece of wood, the

more difficult it is to get it to burn. Although this subject is covered on page 79, an example from "Textbook of Wood Technology" (8), follows. When during a fire, unprotected steel beams 60 feet long (supporting a floor) are heated to 1100°F. they elongate nearly 5 inches. If this doesn't cause collapse of the walls, the building will soon come down anyway because at this and higher temperatures (1700° or more) the steel becomes soft and cannot even support its own weight. Heavy timber construction using large beams is much safer since wood in this case burns slowly and except for a burning layer outside, the interior is unaffected and still strong.

The fact that wood decays, and also burns, is in many cases an advantage. If fallen trees in the forest never decayed, they would finally cover the ground to a depth that would make new growth impossible. Wooden objects including torn down houses no longer wanted can be disposed of by burning and, if dry, there is little smoke to pollute the air. Metal objects (especially iron or steel) currently including old automobiles, pile up in unsightly "graveyards" and require many years to rust away. At present it does not pay to re-use all such waste metal. Hopefully, in the future, we shall need to use it as mined metals become scarcer.

(4) Wood may be attacked by various borers or carpenter ants that tunnel in it. In manufactured articles such trouble is rare. Ants usually do not attack dry wood. Again, as with decay, wood must be kept dry. Marine borers (Teredo) have always been a problem where wooden piling is used in salt water. Heavy impregnation with creosote and tar repels them. Important in warm to hot climates are termites which mine out wood leaving a thin outer shell that collapses when weight is put on the mined beam or object. You might sit on a mined chair and have it collapse under you. House protection by termite shields or by poisoning the soil are two methods of control.

(5) Compared to the metals and synthetic materials, wood is variable in its properties. The great variation between species is not too much of a problem provided one knows the features and can specify the kind of wood to be used. Variation within a species may cause some trouble and is much harder to determine since it depends upon a number of factors such as ring width, amount of latewood in each ring, and the dry weight of the wood. Except in extreme cases, this kind of variability is usually ignored but can never be entirely forgotten.

Some of the more obvious advantages of wood are: (1) It can be worked either with simple hand tools which have not changed much during the last thousand years, or with the newest of power-driven labor-saving machines. (2) Strong joints can be produced using nails, screws, bolts, or special timber connectors, all of these applied by hand, or powered tools. (3) New adhesives far more efficient than the traditional glues are readily applied to wood to form exceedingly strong and lasting bonds. As an example, large laminated beams support roofs with a clear span of 200 to 300 feet or more. (4) Considering its great strength, wood structural members are relatively light in weight and are easily handled either in manufacturing or on the construction job. (5) Wood resists corrosion in contact with a large number of chemical solutions, and so it is used for vats or pipes to store and transport them. (6) Wood is not only an excellent insulator against extremes of heat and cold, but it is the only commonly used thermal insulator that also has good structural properties. (7) Dry wood is a poor conductor of electricity. Wooden poles, cross-arms, and insulator pins contribute insulation. Wooden ladders are safer than metal ones in fighting fires where broken power lines may be dangling. A dry board may be used to flip a live wire away from someone upon whom it has fallen. (8) Wood absorbs energy much better than concrete. This is why walking on a wooden floor is less tiring than on one of concrete. Railroad ties of wood are easier on the rolling stock than those made of concrete. Watch a moving train as the wheel trucks momentarily depress the wooden ties and how they then spring back into place. (9) For interior paneling and furniture in homes and offices nothing approaches wood for warmth and beauty of grain and texture. Plywood panels of unending diversity of pattern soften the hard lines of steel and concrete in modern buildings. Wood furni-

Figure 82 (left to right Shrinkage of tangential, radial and lengthwise cut pieces of oak wood; and plywood

Figure 83 Result of tangential shrinkage in an air dried cross section of a log

ture because of its poor heat conductance never feels too hot or too cold as metal furniture often does. Walk across a rug and reach out to touch a metal filing case and you often receive an annoying shock of static electricity. This never happens with wood.

Several other important advantages of wood have been mentioned at the beginning of this chapter where its less desirable features were discussed. This short list by no means covers all of the unique features of wood, and by observing as well as looking about you, a much longer one can be made.

The principal uses of wood are (1) as a structural material for building and decoration, (2) for pulp to make paper and other wood pulp products where it reigns supreme in the amounts used, (3) as a source of chemicals of great number and diversity, with a potential for many more, (4) as fuel wood. In advanced societies, wood as fuel is now mostly used only as a luxury in fireplaces. Charcoal is of course used in enormous quantities in this country for backyard cooking. Primitive societies have in many cases burned for warmth and cooking most of the available wood. Now, in the Old World, great areas originally forested are barren. Attempts at re-forestation often fail because of the ubiquitous goat that hungrily eats anything green that pokes above the ground.

Moisture, Swelling and Shrinking

We have seen that wood is a marvelous and beautiful structure of innumerable hollow cells. Wood properties depend upon the relative thickness of the cell walls, their chemistry, and the way in which the various cell types are arranged to form the woody structure.

A piece of dry wood takes up moisture in both vapor and liquid forms. The temperature and humidity of the surrounding air determine how much water vapor the wood will attract and hold. The humidity of the air is constantly changing, and this affects the amount of moisture in the wood which the air surrounds. From the viewpoint of use, we might perhaps "forget" all this except for one thing. When a piece of dry wood absorbs water vapor from the air, the wood gets *larger,* and a piece of wet wood surrounded by dry air gets *smaller*. Wood, then, is *constantly changing in size* following changes in atmospheric humidity. This feature of wood must be understood to use it properly. The general principle is to dry wood to a moisture content in balance with the humidity where it is to be used. Before further discussing size changes in wood, it will help to know how to determine the amount of moisture in a piece of wood, the location of such moisture, and how it moves through wood.

To find the moisture content (M.C.) of a piece of wood, it is first weighed. This figure includes the weight of the wood plus that of the water it contains. The piece is then oven dried at a temperature of about 216°F (3 to 4 degrees above boiling), until the weight is constant. This is the oven dry (OD) weight of the wood itself. The OD weight is subtracted from the first figure (wood plus water) to give the weight of the water in the piece of wood. A useful formula is

$$\text{M.C.\%} = \frac{\text{Weight of wood and moisture minus OD weight of wood}}{\text{OD weight of wood}} \times 100$$

Water in wood may be found both in the cell walls themselves, and also in the vast network of cell cavities. The forces of attraction between dry wood and moisture are enormous, so much so that the wood cannot be prevented from taking up moisture from the air. Water molecules enter the highly complex ultramicroscopic structure of the cell walls which swell as the water gets between microfibrils and into yet smaller spaces in the molecular framework. The water molecules are attached to the wood substance by hydrogen bonding, a process that gives off energy. This energy can be measured as the *heat of wetting*. To 100 cc of water add 20 gm of dried sawdust stirred in with a thermometer. The temperature will rise several degrees. As might be expected, to dry wet lumber requires an equivalent amount of heat energy to break the hydrogen bonds between the wood and the water.

The continued introduction of water molecules into dry wood swells it still further, but increasing forces develop against the taking in of more water. Eventually, the forces of attraction and those of swelling balance each other, the cell walls are saturated and can absorb no more water. The moisture content at this stage is called the *fiber saturation point.* Above the fiber saturation point, there is no more absorption of water in the cell walls and no more swelling. The volume of a piece of wood at or above the fiber saturation point is called the "green volume" and is a useful quantity to use since it is a constant. Although there is some variation between species, the moisture content is usually about 25 percent at fiber saturation.

In the above discussion it is assumed that water has been absorbed by the cell walls, but that there is *no liquid water* in the cell cavities. If a piece of water-vapor saturated wood is now placed in water, the cell cavities begin to fill as air is displaced. This can be hastened by forcing the wood under water, or by putting the container in a vacuum chamber. Finally, the wood will be completely "water-logged" with all the cell cavities filled. This is the point of highest moisture content. Structure, and especially relative cell wall thickness, determine how much water a wood can hold. Light weight woods with thin cell walls and large cavities can obviously hold the most water; their maximum moisture content may be 200 percent or more. Heavy woods with thick cell walls and small cavities may hold a maximum of 60 percent or less.

Great variations may be expected in green lumber just cut from the tree. In coniferous trees, the sapwood is much wetter than the heartwood. As an example from the "Wood Handbook" (13), green sugar pine sapwood showed a moisture content of 219 percent; the heartwood registered only 98 percent. A different condition is found in the broad-leaved trees. Here, the variation between sapwood and heartwood may be small, and also the heartwood may be the wetter. A piece of hickory showed the moisture content of its heartwood to be 70 percent, the sapwood only 52 percent.

The movement of moisture through wood is influenced greatly by its tubular structure with the tubes running lengthwise in the tree. Moisture may move 12 to 15 times faster along the grain than it does across it. In a cube of wood, most of the water evaporates from the two end surfaces. However, in a board one inch thick and 8 to 10 feet long, more moisture will pass into the air across the grain because the distance (max. one half inch) is so small compared to the length.

When a "green" or wet piece of wood begins to lose moisture to the air (especially on a hot day) the outside layers may become fairly dry, while the inner core of the piece is still quite wet. The outer layers shrink as they dry, but cannot do so fully because of the wet core. They "set" partially shrunk, and when the core finally dries, the piece of wood is left with internal stresses. The relieving of these stresses, and the whole subject of air drying and kiln drying of wood is beyond the scope of this book. "Textbook of Wood Technology" (8), the "Wood Handbook" (13) and other publications cover this very important and technical aspect of wood-moisture relations. In air dried lumber stored outdoors under cover, the moisture content is usually between 12 and 15 percent. In heated buildings, the amount of moisture held by wood is much lower, and may vary from about 10 percent in summer to as low as 4 percent in winter.

We have already seen that at the fiber saturation point, the volume of a piece of wood is at a maximum. It is as large as it can be from absorbing water. If drying now begins, the wood structure starts to shrink, but *not evenly* in all three directions. Lengthwise, shrinkage is very small, usually not more than 0.1 to 0.3 percent from the saturated to the oven dry condition. Shrinkage across the grain is much greater. Radially it may be from 3 to 5 percent (depending upon the kind of wood). Tangentially, shrinkage is nearly *twice* the radial shrinkage. These figures may be very important when applied to house or other construction. Green or partly dried wood must not be fabricated where the moisture content of the air is low, and will remain so. If this advice is not followed, do not be surprised if cracks develop between boards originally nailed tightly edge to edge.

Figure 82 shows the relative shrinkage to oven dryness in the three directions of oak wood. When green, the three pieces had the same height. They were cut so that one piece was lengthwise of the grain, the second radial, and the third tangential. For comparison, a piece of plywood stands beside the one with the grain running lengthwise. In plywood, three or more sheets of wood veneer are glued together with the grain direction alternating at right angles. The resulting sheet of plywood shows little shrinking or swelling in any direction across its surface.

Another example of the radial-tangential shrinkage effect is shown in the radial checking or splitting of a log cross section as it dries (Fig. 83). To prevent or modify this, soak the green section for between one and four weeks (with a weight on it to make it sink) in a 50-50 solution of polyethylene glycol 1000 in water.

Specific Gravity and Buoyancy

Specific gravity is a measure of how much a piece of wood (or other material) weighs in comparison with the weight of an equal volume of water.

$$\text{sp. gr.} = \frac{\text{ovendry weight of wood block}}{\text{weight of same volume of water}}$$

Using 62.4 lbs. as the weight of a cubic foot of water, suppose that a cubic foot of ovendry wood weighs 31.2 lbs. Then $\frac{31.2}{62.4} = .50$ the specific gravity. Wherever you see the sp. gr. for a material listed, its weight per cubic foot is found by multiplying the sp. gr. by 62.4.

When determining the sp. gr. of a wood block it is not at all convenient to actually use a perfect cube one foot on a side. Much smaller blocks are used, and their dimensions need not be exact provided we use the principle of *displacement*. Suppose we have a block of the same kind of wood used above. Long ago, Archimedes discovered that a floating body displaces its own weight of water. As we place the block in water, the water rises all around the inner surface of the container. Now we must get the *weight* of this displaced water (there are several ways of doing it). Then

$$\text{sp. gr.} = \frac{\text{ovendry weight of wood block}}{\text{weight of water displaced}}$$

We already know that the dimensions of a dry wood block increase as it takes up moisture. When ovendry, the block is smallest; when at the fiber saturation point it is the largest it can be and additional liquid water in the cell cavities causes no further swelling and increase in volume. For the same block of wood, then, there are two extremes in size, one when it is ovendry, the other when it is swollen to the fiber saturation point. This "green volume" is the more constant to use in determining specific gravity. So now we have finally

$$\text{sp. gr.} = \frac{\text{ovendry weight of wood block}}{\text{weight of water displaced by the block at or above the fiber saturation point (green volume)}}$$

Of course, either the volume when ovendry, or the volume at the fiber saturation point may be used, but the values obviously will be somewhat different. In tables, always look to see which volume was used. Some tables give both values, or even values based upon the volume at some intermediate moisture content.

Since a floating body displaces only its own weight in water, it has a certain buoyant capability left over. If a cubic foot of wood weighed 62.4 lbs., it would have no buoyancy whatever, but would float with its upper face approximately at water level. If one could imagine a weightless cubic foot of wood, then one could load it with 62.4 lbs. before it would sink! From this it can be deduced that if a floating cubic foot of wood sinks halfway, then its weight is 31.2 lbs. and its buoyant force is also 31.2 lbs. If it sinks ¾ of its vertical dimension, then ¼ of it is left above water. One quarter or 25% × 62.4 = 15.6 lbs. which can be loaded upon it to bring it to the sinking point.

Suppose a cubic foot of white pine wood has a sp. gr. of .34 based on green volume, and a moisture content of 70 percent. How many pounds load will it carry in water before sinking? First we find its dry weight by multiplying .34 × 62.4 = 21.2 lbs. Then to find the weight of the included water multiply .70 × 21.2 lbs. = 14.8

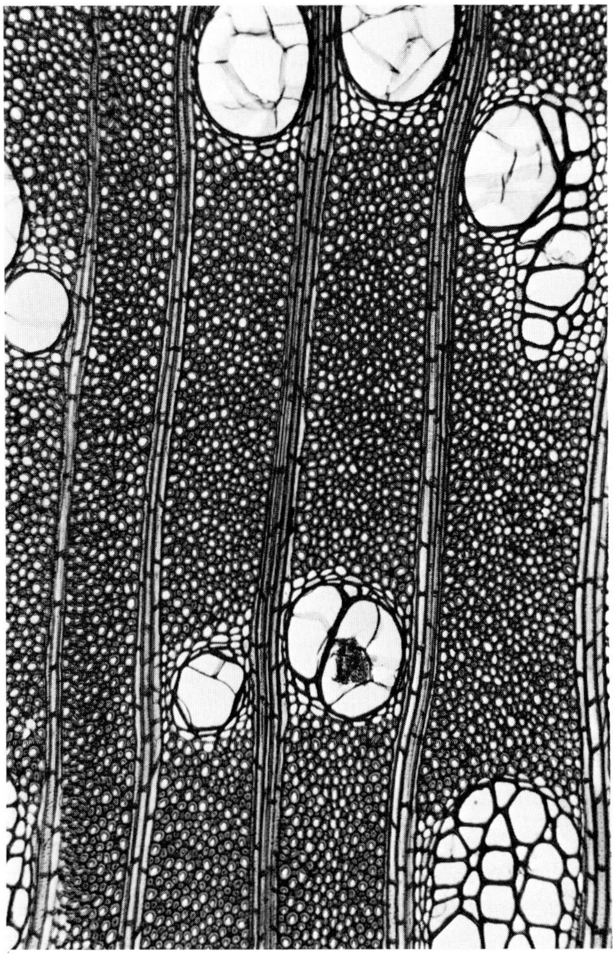

Figure 84 Black locust × 75 Sp. gr. .66

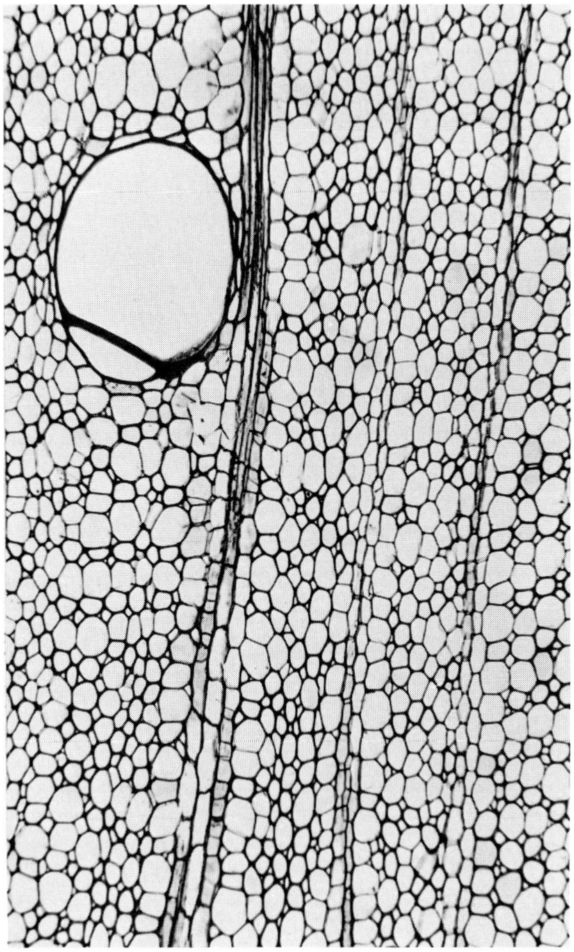

Figure 85 Balsa × 75 Sp. gr. .12

lbs. Now add 21.2 lbs. (weight of wood) to 14.8 lbs. (weight of water) = 36 lbs. total weight. Subtract this from 62.4, and you have the buoyant force which is equivalent to 26.4 lbs. To find how deep in water the 36 lb. block will float, $\frac{36}{62.4}$ = approximately 58% submerged.

These two woods represent extremes in weight. The black locust is more than 5 times as heavy as the balsa (a famous wood from South America). Even a glance at the two illustrations is enough to see that one is heavy, the other light.

In cross section, the fibrous cells that make up most of the wood structure in black locust are very thick-walled with small cavities. Most of the wood is cell wall substance with but little air space throughout. By contrast, the wood of balsa is mostly air space, and the cell walls are extremely thin. Balsa can be translated as "raft", and logs of it are used to buoy other green hardwood logs that would otherwise sink. Such rafts are run down rivers to the coast where both the heavy hardwoods and the balsa are sold and loaded aboard ships to begin their journey to ports of the world.

Some woods (mostly tropical ones) are much heavier than black locust, and certain others are even lighter than balsa. Among heavy ones are leadwood, lignumvitae, and purpleheart. Even when oven dried all of these woods sink in water. Wood substance itself without any air spaces is about 1½ times heavier than water. Therefore the weight and hence the buoyancy of a piece of dry wood depends upon the volume of the air spaces enclosed by the billions of cells of which it is made.

The Chemistry of Wood*

Chemical analysis in general deals with such things as substances in solution, solid precipitates that can be weighed and identified, distillation and the separation of its products, boiling points and freezing points, and many other more sophisticated techniques for separating a sample of material into simpler identifiable portions. As an example, if you mix together solutions of salt (sodium chloride) and silver nitrate, a white precipitate forms immediately. This is silver chloride which can be dried, weighed and the amount of silver calculated.

$$NaCl + AgNO_3 \rightarrow AgCl + NaNO_3$$

(*water solutions*) (*solid*) (*in solution*)

This reaction is especially interesting because it is the basis for the great photographic industry which uses enormous amounts of silver salts in producing films and papers. Silver chloride, bromide, and iodide are extremely sensitive to light and in suspensions in gelatin form the emulsion on photo materials.

Early in the nineteenth century, wood was thought to be a uniform chemical compound, and the first analysis showed that it was composed entirely of carbon, hydrogen, and oxygen. From this simple beginning up to the present, hundreds if not thousands of organic chemists have discovered that wood is not only a heterogeneous structure of the greatest imaginable chemical complexity, but also that it strongly resists the usual methods of chemical analysis. Drastic ways are usually required to take it apart, and this means that the original wood substances may undergo considerable change during analysis. Although much has been learned, modern chemists still find wood chemistry a puzzling but challenging labyrinth to explore. Because of the length of time needed for an analysis, very few of the many thousands of kinds of wood have been investigated, and a man might spend his lifetime studying the chemistry of a single species.

* Whether it is possible to present this subject without presuming a year's study of general chemistry and another of organic chemistry might well be debated. It is hoped this discussion will have some meaning even to those whose knowledge of chemistry is slight.

We have seen that wood is a marvelous structure made up from an almost infinite number of cells each with its own cell wall. For many uses, it is important to know where in the cell wall various substances are located. But in wood analysis the sample is mechanically reduced to a powder with the particles of an arbitrary size (i.e. those that will pass a screen with 40 wires to the inch and be retained on one of 60 wires). Coarser, and also finer particles are in this way eliminated and the sample has approximately the same sized particles throughout. This may be important when several workers try to compare their results on the same kind of wood. Obviously, wood structure must be ignored, and analysis proceeds using methods already developed in other branches of chemistry.

Many kinds of wood (especially the heartwood) are permeated with chemical substances known as extraneous components. A wood is often characterized by a certain substance or group of them not found in other woods. One or more of these may give great resistance to the attacks of wood destroying fungi. Once upon a time a giant redwood fell to earth, and sometime later, on top of the prostrate trunk, a tiny redwood seed germinated. In the cool moist forest, the little tree kept growing until finally after many years, its roots reached the ground on both sides. Hundreds of years passed and the tree reared its great trunk over the old tree. When the towering tree was finally logged it was nearly a thousand years old. Meantime the sapwood of the prostrate trunk had decayed and disappeared, but the heartwood was still as sound as the day it fell, probably more than a thousand years before the tree that grew upon it was cut. It is presumed that the heavy impregnation of redwood heartwood with tannins and perhaps other extraneous materials makes it so durable. Other trees such as the cedars have heartwood that is also unbelievably resistant to decay. Here various oils promote durability.

Besides resistance to decay, some other or the same extraneous substances may affect or determine certain physical properties of wood. These include odor, taste, color, inflammability, toxicity, and even density. In quebracho wood from

South America, and a few others, tannin may comprise one quarter of the weight of the dry heartwood. Each of the native cedars has its own distinctive odor given off by the oils in the heartwood.

Extraneous materials in wood number in the thousands. "Wood Chemistry" (18) devotes some 145 pages to these substances alone, apart from the chemistry of the wood material itself, and lists more than 500 scientific papers on them.

Tannins, mentioned above, are widespread in nature and are often found not only in the wood, but also in some other or all parts of a plant. They react in a very complex not well understood way to transform raw animal skins into leather. Certain tannins give a blue color when treated with iron salts, others turn green. Common sources have been hemlock bark, oak bark, and sumac leaves. Presently, most commercial tannin comes from quebracho wood.

Another class of extraneous substances includes the oleoresins produced in special canals in the wood of many kinds of conifers and a few hardwoods. The coniferous resins yield turpentine, tars, and rosin, the so-called "naval stores". Among the many volatile oils from certain species should be mentioned those from cedar wood, sandal wood, sassafras, and the camphor tree (all parts). Camphor oil has some 30 different constituents.

The heartwood of most trees is darker than the sapwood. At the time sapwood becomes heartwood, various colored substances (besides tannins) are deposited. Osage-orange chips yield a colored solution when boiled in water. Some of these colored materials have yielded important dyes which, over the years, resist fading much better than the cheaper man-made aniline products of today. "Logwood" from southern Mexico and Central America yields haematoxylin from which a bluish-black dye is made. The fabric is first soaked in a mordant (iron-alum salts) which becomes affixed to the fibers. Then after the fabric is washed, the oxidized haematoxylin (hematein) solution is added and the hematein and mordant combine to produce a fast color. Haematoxylin is still one of the very best stains for staining plant and animal cells to be viewed with a microscope. The illustrations of cell structure in this book are mostly from sections in which the middle lamella was stained with haematoxylin. Certain "Brazil Woods" when first cut are yellow in color, but soon turn red when exposed to the air or oxidizing reagents. The red brazilein is an excellent dye, and the earliest settlers along the coast of Brazil came there to cut dyewoods long before artificial dyes were known.

The extraneous substances in wood are also called extractives and before proceeding with the analysis of the wood itself, they must be removed. This is usually done by successive hot extractions of the sample with an alcohol-benzene mixture, alcohol, and water. Although this suffices with most woods, some of them are very difficult to free completely of all their extraneous components.

Before entering the world of the molecule, we must attempt to understand the sizes of things. Man is "middle-sized" in his relation to the universe. According to Sir Arthur Eddington, about 100,000,000,000,000,000,000,000,000,000 (10^{28}) human bodies would equal the volume of a star (a small one, perhaps!) while about 10,000,000,000,000,000,000,000,000,000 (10^{27}) atoms constitute a man. There are about 25 millimeters in one inch. The unit for use in the molecular range of size is the Angstrom (Å). One Å equals one *ten millionth* of a millimeter.

To complete our discussion of the cellulose model (Fig. 86), the detached upper cellobiose unit acquires on its bottom carbon atom an -OH group and in the unit below, the top oxygen gains an -H (hydrogen) atom. Such a reaction is called hydrolysis in which a molecule of water H-O-H (H_2O) is added. Now, each of the two units is cellobiose ("wood sugar") with exactly the same number and kind of atoms that constitute ordinary cane sugar, sucrose, $C_{12}H_{22}O_{11}$. The difference between the two lies in the *arrangement* of the atoms in the molecular structure. Having already noticed that there are four units in the cellulose molecule connected by oxygen bridges, it seems logical that by hydrolysis each molecule of cellobiose will be split in half at its oxygen bridge and by the addition of H-O-H two

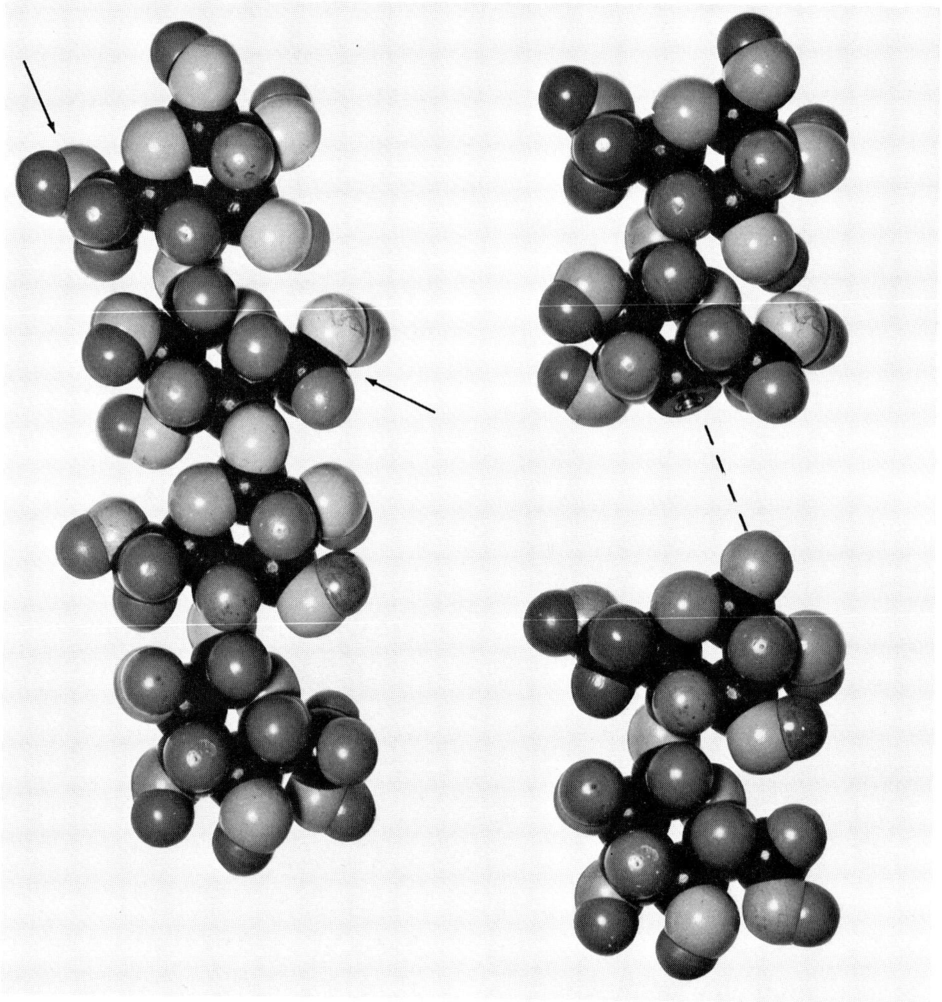

Figure 86 Scale model of a much shortened cellulose chain molecule. The width is about 3 Angstroms. The jet black spheres are carbon atoms, a few of which are hidden in the illustration. The lightest grey atoms are oxygen, and the next darker ones are hydrogen. Two of these together represent an OH group. Run your eye from top to bottom and notice that there are four clusters or groups connected by oxygen atoms or bridges. Now notice that there is an alternating symmetry in the molecule. At the left of the top unit a barely visible carbon atom is holding an OH group and two hydrogen atoms (CH_2OH). In the next unit below, this group is at the right, and the carbon atom is plainly visible, but one of the hydrogen atoms barely shows.
At the right, the cellulose molecule has been broken to show two cellobiose units, the repetitive unit in cellulose long chain molecules.

molecules of a simple sugar will result. This indeed happens, and the simple sugar is the widely used glucose. Actually, during hydrolysis with strong acids all four oxygen bridges are broken and the final product is glucose. Special methods are needed to prepare cellobiose.

It has already been stated that the model represents a "much shortened cellulose chain molecule". This is seen to be a gross understatement when it is realized that there are about 10,000 glucose residues or groups in an average cellulose long chain molecule! If one could travel the length of a cellulose chain molecule, he would find that its space relations with nearby cellulose molecules is changeable. For a certain distance, adjacent chains maintain an exact spacing with their neighbors and are bound together in a crystal structure which transmits polarized light and shows other properties of crystals. Then as one moves along the bundle of cellulose chains forming the crystal, they lose their alignment, the spacing becomes random and crystallinity ceases. This amorphous zone soon gives way to another crystalline region, and so on through the length of the chain molecule. In cellulose from wood pulp, from 60 to 70 percent of the cellulose is in the crystal form.

When wood cellulose fibers (page 26) are treated chemically they break down into thread-like microfibrils of smaller and smaller size. For many years microscopists have tried to determine the size of a "unit fibril", the basic building unit of the cell wall. The light microscope gave a greatly exaggerated size to such units. The electron microscope, with its great leap forward in resolving power, now indicates that the smallest aggregate of cellulose chains is the "elementary fibril" with a diameter of 35 Å. Each elementary fibril consists of about 40 cellulose chains. Recalling that the breadth of the chain is 3 Å, it is clear that within the elementary fibril there is considerable *space* between the chains. The importance of this space will soon be seen. Cellulose is widely distributed throughout the plant kingdom, not only as a principal constituent of cell walls, but also in such things as seed hairs. Those of cotton are almost pure cellulose. Wherever cellulose has been analyzed, be it in a moss, a fern, an herb or a tree, it has always been found to have the structure shown in the scale model.

Besides cellulose, there are present in the woody cell wall two classes of amorphous substances, the hemicelluloses, and lignin. These, presumably along with water, fill the relatively large spaces between the cellulose chains and also those between fibrils of various sizes. The hemicelluloses also have chain molecules of simple sugar units, but their length (150 to 200 units) is many times less than that of cellulose. To the chemist, the hemicelluloses present a most interesting and varied group upon which much research has been done. To say more about them is beyond the province of this discussion which presupposes little or no knowledge of organic chemistry.

The subject of lignin is enormous as well as of the greatest importance, especially to those industries which use cellulose fibers, where the lignin in the form of derivatives is a waste product and stream pollutant. The chemical structure of softwood lignin has defied attempts at discovery for many years, but seems now to be solved, or nearly so. It has a very complex molecule featured by a 6-carbon benzene ring from which extends a straight chain of three carbon atoms with -H atoms, and various groups attached. A methoxy group, $-CH_3O$, and a phenolic -OH group are attached to the benzene ring. Such a structure is called "aromatic". A number of these relatively simple units are combined by cross links to form the complex lignin molecule. During the process, some alcoholic -OH groups are added. Lignin of the hardwoods is much more variable and complex than that of the softwoods, but the fundamental unit appears to be the same except for the addition of another methoxy group on the benzene ring in the hardwoods. The formulas for the two types are shown on page 76. That on the left features the softwoods, while *both* types are found in the hardwoods.

The diagram below summarizes our discussion to this point. Notice that the cellulose plus the hemicellulose is called holocellulose (whole cellulose).

|—— holocellulose	——|	
Cellulose	**Hemicelluloses**	**Lignin**
long-chain molecules →	short chains containing galactose, arabinose, mannose, xylose and uronic acid units; methoxy, and acetyl groups	consists of the non-carbohydrate aromatic materials of the cell wall.

Diagram showing the composition of extractive-free, ash-free wood.

Cellulose and lignin, and to a lesser extent the hemicelluloses, are bound together in a cell wall structure of great mechanical strength, possessing

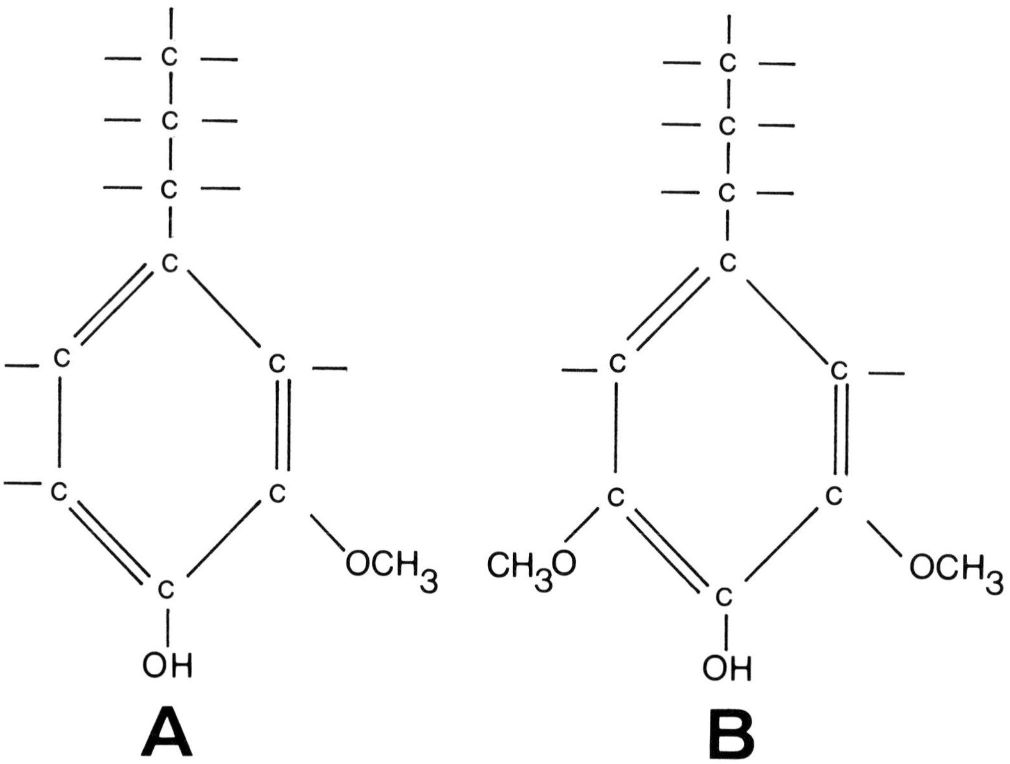

Basic units in the lignin molecule, A, softwood lignin. B, hardwood lignin. The vacant bonds indicate where other groups, or hydrogen atoms are attached in the building of the complex lignin molecule from these relatively simple units.

as well strong resistance to the usual chemical reagents. There is perhaps no chemical available for extracting one of the three components that does not also degrade it, as well as affecting the two remaining ones. Even the boiling of wood in water for a week or more alters, even though it may be slightly, its chemistry.

Through the years, certain standard methods have been developed for wood analysis, and these are given in "Testing Methods and Recommended Practices" of the Technical Association of the Pulp and Paper Industry (TAPPI). Only enough of the methods is given here to indicate the chemicals used and something of their effects. These are all experimental, arbitrary procedures, and their value lies in various researchers and technicians being able to obtain comparable results. It is assumed that the samples of extractive-free wood were prepared as previously indicated. Various washings with water during treatment are omitted.

Cellulose. The purpose is to remove the lignin only, but some of the hemicellulose portion is also removed, and even the cellulose may be altered. Chlorine gas is passed through the moist sample, followed by washing with weak solutions of sulfur dioxide, and then sodium sulfite. The treatment is repeated until the sample is white or nearly so.

Holocellulose. According to TAPPI, "Holocellulose is the lignin-free fibrous material comprising all of the hemicellulose and cellulose in wood. It is white, cream, or straw-colored depending upon the kind of wood". Holocellulose is prepared by passing chlorine gas through the sample followed by a hot dilute solution of

Chemical Composition of Wood*

(All values in percent of extractive-free wood)

Softwoods (conifers)

Component	Balsam fir	White spruce	Eastern white pine	Eastern hemlock	Northern white-cedar
Cellulose	42	41	41	41	41
Hemicelluloses	27	31	27	23	26
Lignin	29	27	29	33	31
Pectin, starch, ash, etc.	2	1	3	3	2

Hardwoods (broad-leaved trees)

Component	Red maple	White birch	American beech	Quaking aspen	American elm
Cellulose	45	42	45	48	51
Hemicelluloses	29	38	29	27	23
Lignin	24	19	22	21	24
Pectin, starch, ash, etc.	2	1	4	4	2

These tables are included to show the variation, and also the relative uniformity in wood components between softwoods and hardwoods and various species in each group. Although no two woods are the same, they do show a similar pattern.

*Analysis by Tore E. Timell, adapted from similar tables in "Principles of Wood Science & Technology" Vol. 1. Kollmann, F. F. P. & W. A. Côté, Jr., 1968.

monoethanolamine in alcohol. The cycle is repeated until the residue is white, or fails to change color any further.

Lignin is so tightly bound to the cellulose skeleton that for many years they were referred to as "lignocellulose", implying chemical bonds between the two. This idea has been abandoned, but it does suggest the extreme difficulty of separating the two without changing them chemically. The TAPPI procedure for isolating lignin involves a very drastic treatment of the wood sample with 72% sulfuric acid. This removes by hydrolysis the holocellulose, and leaves the lignin, somewhat degraded, as a residue. After thoroughly mixing the acid with the wood sample, the mixture stands, with periodic stirrings, for two hours. It is then diluted and boiled to complete the hydrolysis. Lignin to be used for research on its molecular structure is obtained in solution using a much less drastic method.

Microchemistry of Wood

At the beginning of our discussion of wood chemistry, it was made clear that the samples used were finely ground wood treated with various chemical reagents. The results of course could have no meaning in terms of wood structure or the location of the several components (i.e. cellulose, lignin, etc.) within this structure. Chemists are not trained to make thin sections of wood and to observe them with a microscope while various chemicals are applied. Few botanists practicing the art of wood sectioning have also had the chemical background to explore such a field, and anyway most botanists have considered wood a "refractory substance" sure to turn the edge of their super sharp knives! The result has been that with some exceptions, botanists much prefer to work with soft tissues that need no further softening, and can be embedded in a small block of paraffin for sectioning. Within recent decades the wood technologist has come to the scene equipped with both botanical and chemical methods to locate the several principal components of the cell wall. However, it should be said that a few chemists and a few botanists have made and are making outstanding studies in wood microchemistry. At this point it

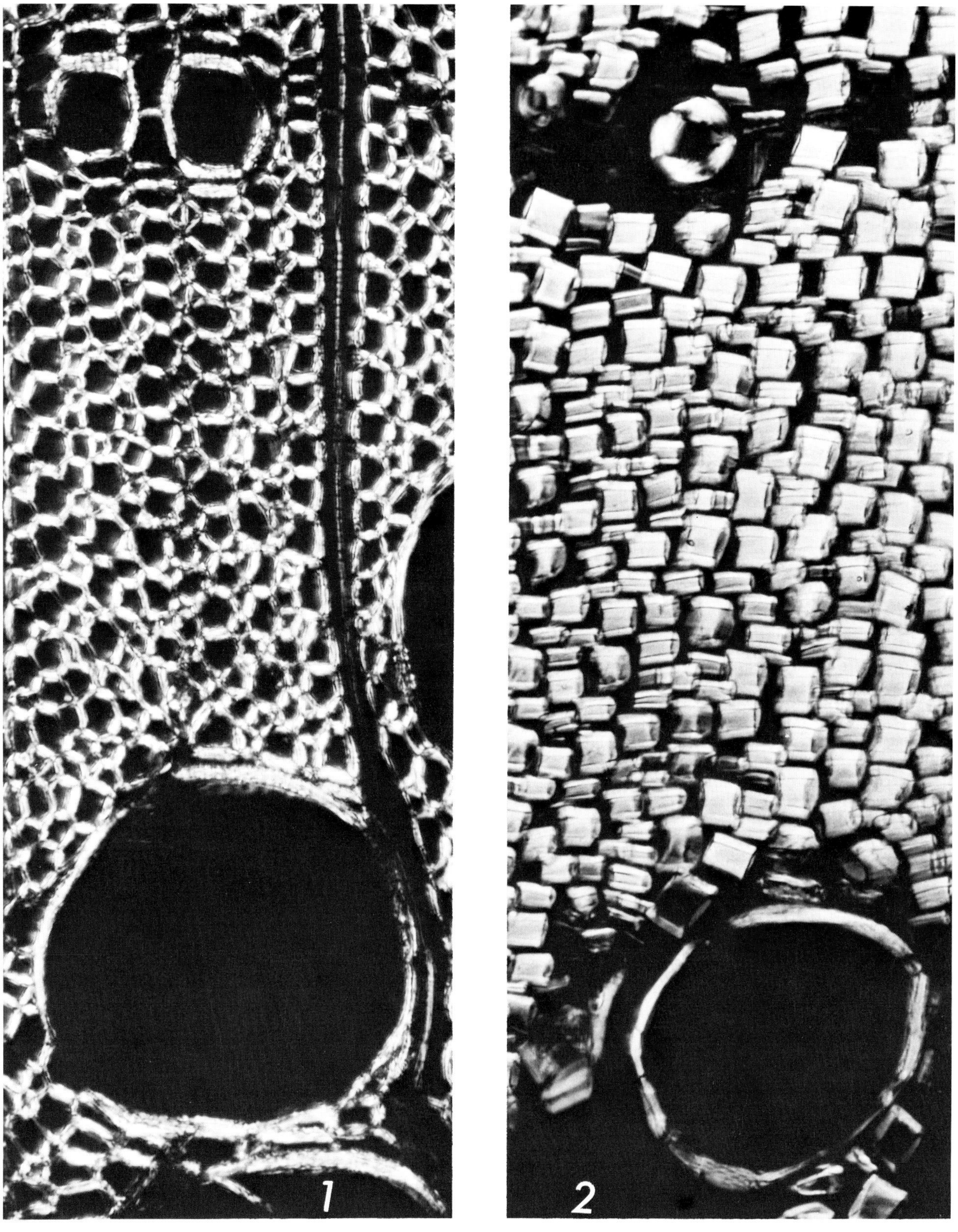

Figure 87 Black Ash cross-sections viewed by polarized light, × 600. No. 1 untreated No. 2 delignified; the separated fiber cells are lying on their sides (see text).

might be well to look again at the model of a wood fiber section on page 27, and also the middle lamella network on page 30. The fiber section shows the primary wall on the outside, and as in paper fibers in general, the middle lamella or intercellular substance has been removed by some process of delignification.

The illustrations of thin sections of black ash (Fig. 87), one untreated and the other delignified, show that complete removal of the lignin allows the fibers to separate; it is as though the middle lamella binds the fibers and other cells together and when it is removed they either fall apart, or upon slight disturbance do so. The photo of the delignified section is unusual in that the separate fiber segments are lying on their sides; the length of these segments is the thickness of the thin wood section treated. This special effect was produced by moving the cover glass slightly. When first mounted in a drop of water, the cells were all upright. It takes very careful handling to treat such a thin section to remove the lignin without having the cells separate prematurely. Consider now, that although the lignin comprising about 25 percent of the weight of the wood has been removed, the fiber segments appear whole. Since visually, only the middle lamella has gone, you might think that the fiber walls have no lignin but this would be wrong. By other techniques, instead of removing the lignin from a wood section, one may remove the holocellulose leaving an intact lignin skeleton. In such a section, the fine, extremely porous structure of the cell wall remains, together with the dense supporting network of the middle lamella. It is now clear that one may remove either lignin, or holocellulose and still leave behind a coherent structure of the cell wall. These two interpenetrating systems have been compared with reinforced concrete, where lignin is the cement and the cellulose long chain molecules and fibrils are the reinforcing rods. In any case the woody structure has enormous strength for its weight as we shall see presently.

To complete this short discussion, it should be said that the term middle lamella has been used so far to denote the true intercellular substance present originally in the cambial region. The cambial cell walls are the primary walls of mature woody cells. The 3-layered secondary walls are laid down within the primary walls (model p.27). As in other soft tissues, the cambial middle lamella is largely of pectin materials (the jellying principle in fruits) but as the woody cells mature, the middle lamella becomes mostly lignin, and these materials also heavily impregnate the flanking primary walls, and usually to a lesser degree the secondary walls.

In mature woody cells it is extremely difficult or impossible with a microscope to see on each side of the middle lamella the primary wall. With the tremendous magnification of the electron microscope, even though ultra-fine structural details are clearly seen, it is still hardly possible to see where the true middle lamella ends and the primary walls begin. For this reason, the term *compound middle lamella* is used to denote the three layers taken together, primary wall, middle lamella, primary wall. All of the photomicrographs in this book show just the compound middle lamella, and the secondary cell walls.

Reaction to Heat and Fire

When wood and also other materials are heated, their molecules move or vibrate faster, and the distances between them increase. This makes the entire piece of material larger. Cooling has the opposite effect. The molecules slow down, and the distances between them decrease. In wood, moderate temperature changes affect the size of a timber very little. If you cool a dry 16 ft. plank of white ash from 80°F to -20°F it will only be about ⅛ of an inch shorter at the lower temperature (a difference of 100°F). Across the grain, the change in dimension may be from 2 to 3 times as much. A floor 16 ft. wide made of tightly fitted pieces might shrink across the grain ¼ to ⅜ of an inch or more, depending upon the kind (species) of wood used. These figures assume that there is no moisture in the wood, but in practical use, there is always some moisture present. As wood is heated, the moisture is driven off, and the piece shrinks. This tends to offset the slight swelling produced by heating and so wood is more stable dimensionally than it might appear.

The amazing heat insulation provided by wood with its billions of air filled tubes, fibers, and brick-shaped cells is not appreciated by most people. To equal the heat insulation of a redwood wall 4 inches in thickness, one would need a brick wall two feet thick, or a concrete one (gravel and sand aggregate) five feet thick. A common example of the poor heat conductance of wood is the behavior of a wooden or paper match when lighted. The match end between thumb and finger stays cool as the flame comes nearer and nearer. Finally, when one blows out the flame to prevent being burned, the stub end of the match is still cool. Try this with a piece of wire of the same length, the tip held in a flame. In a matter of seconds the wire will be too hot to hold.

The art of producing fire was man's first great discovery, now lost in the mists of time. It gave him immediate superiority over the animals, and blazing torches not only kept them away at night, but also partially lighted his cave, while it cast mysterious dancing shadows on the walls.

Watching a wood fire burning in a fireplace either indoors or out is one of the most primitive and soul satisfying experiences of man. The appearance of a piece of wood burning is deceptively simple. Actually the most complex chemical reactions imaginable take place as the wood burns to charcoal and this in turn glows and disappears leaving only ashes— the minerals originally taken from the ground, usually amounting to less than one percent of the weight of the wood. Burn 100 pounds of wood and you will have left less than a pound of ashes. All the rest, the ninety-nine pounds of wood substance came from water brought up from the ground, and carbon dioxide in the air taken in by the green leaves. There, the green chlorophyll, using the sun's energy, builds these raw materials into simple sugars, the basic food of all green plants. In the living cells, hundreds of other organic substances are also synthesized. And now as you watch the flames dance, all of the beautiful wood structure, some of it perhaps formed more than a hundred years ago, is vanishing, never to be seen again. As the fire burns, the wood yields the basic materials from which it was built, water and carbon dioxide, as well as other substances in gaseous form. And the heat it gives off is that of the sun's energy stored so many years ago. When this is all gone, there remains only the gray ashes with their valuable minerals which may slowly sink into the soil and be used in the next cycle of plant growth.

To have a fire as usually defined, there must be (1) fuel in gaseous form, (2) oxygen, and (3) an open flame, or a high enough temperature to cause spontaneous combustion. Remove any one of these three things, and there will be no fire.

The decomposition of wood by burning is called *pyrolysis* and commonly begins when heated wood gives off enough combustible gases to be lighted by an open flame. A thin wood shaving bursts into flame when touched by a lighted match. It might be said that wood itself does not burn. The flames you see are of the burning gases.

The U.S. Forest Products Laboratory, using 1¼″ by 1¼″ by 4″ long samples of dry wood of several kinds, heated them in an electric furnace over a range of temperatures to find out how long it took for them to "catch fire" at each furnace temperature used. A pilot flame was placed one-half inch above the test sample. A portion of their table shows:

Redwood

Temperature	Time for Ignition
356°F	28.5 minutes
482°F	6.0 minutes
806°F	18 seconds

At all three temperatures, some heat passed into the samples, thereby delaying the rise of surface temperatures to the ignition mark. Since wood is such a good heat insulator the process of heat absorption was slow. The higher the furnace temperature, the shorter was the time needed to bring the surface to the ignition point.

These figures are of the greatest importance in considering wood as a safe material for building houses. Even at the lowest temperature given, too hot to support life, one would have more than 28 minutes to get out of a burning building before wooden walls or objects 1¼″ in smallest dimension caught fire. Actually smoke and heat would drive one out much sooner.

When wood "burns", one half to two thirds of the heat given off is from the flaming gases. The remainder is from the glowing charcoal left after the gases are burned away. During the burning gas stage, the charcoal layer having only one third to one half the heat conductivity of the wood, slows down the burning of the layers within.

The poor heat conductance of wood together with the even poorer conductance of the layer of charcoal formed on the outside makes large beams (6″ x 6″) and larger, and wooden columns much more resistant to fire than most people imagine. Although temperatures in a burning building may exceed 1700°F, this figure is often used as a standard in comparing the fire resistance of materials. How will they stand up after one hour at this temperature? Large wooden beams char inwards at about 1½ inches per hour. Furthermore, the transition between the burning charcoal layer without, and the unburned wood within, is quite abrupt. In many cases, when a fire has been put out within a half hour, the charcoal can be planed off and the beam left in place. And meantime, the wooden beam has carried its load and with others has held up the structure. Under similar conditions, unprotected structural steel expands, buckles, twists, and collapses, bringing the structure down in complete ruin (Fig. 88).

The temporary protection of the layer of charcoal which covers burning beams or columns has a number of other interesting aspects. When the Onondaga Indians, and presumably other primitive men, wanted to fell trees for building a stockade, they found that the easiest way was to build a fire at the base of the trunk, and wait for it to burn through. It is recorded that they knew quite well the protective effect of charcoal, and from time to time, knocked it off with their stone axes, exposing unburned wood underneath.

How many times have you seen pictures of a forest of large trees after a fire has swept through it? Did you ever wonder why the trunks were still standing? Even in the hottest fire, after the forest debris on the ground, and the trees' leaves and smaller branches have burned, the great trunks smoulder and go out. In many cases these can be cut and used for lumber. Only the outside is burned and scorched.

You cannot light a dry log in your fireplace with one match or a hundred matches. Blazing kindling and intermediate sized pieces are required to burn the log's surface, and as soon as these are burned up, the fire on the log goes out. Even two logs together burn poorly if at all. Three or more are needed to contribute heat to each other to keep the fire going, in the absence of other smaller pieces.

The California Redwood Association has for many years successfully promoted the use of redwood for constructing *fire walls* four inches or six inches thick in storage buildings and elsewhere. These are commonly built from 2 x 4's nailed together in an upright position. If fire danger is especially high, the finished wall can be surfaced with redwood sheathing. These walls have stood up in a number of fires where unprotected steel collapsed, and that part of the building was destroyed. In steel ships, the transverse, painted bulkheads may not stop fire from spreading throughout the ship. Fire on one side of the bulkhead soon heats it through, and the paint on the other side blisters, smokes and then bursts into flame.

Our own experiments have been with fire resistant boxes carefully made from Douglas-fir and also redwood plywood (both marine grade with no interior spaces), using a resorcinol adhesive which must withstand 1700°F. or more. Papers stored in metal filing cabinets, or cash boxes are cremated in a few minutes, depending upon the speed of temperature rise around them. To protect valuable papers from temperatures climbing rapidly in the first ten minutes to 1300°F with a reading of 1700°F at the end of an hour, a redwood chest may be constructed with an inside volume of one cubic foot, and recessed cover as shown, with the cover, sides, and bottom three inches thick. Metal fastenings for the cover reduce the efficiency of the box. Three vertical dowels glued into sockets on each side of the cover with corresponding dowel holes in the box will keep the cover in place. The weight of such a chest is about 55 pounds. A similar heat-insulated steel box approved by the Underwriters weighs about 90 pounds, and may cost from $40 to $90 or more.

The fuel value of dry wood depends largely upon its weight, although in resinous woods the resins may yield more heat than a like amount of the wood itself. If you look once more at the pictures on pages 40 to 53 it will be clear that the heavier the wood, the more it consists of cell wall substance, with small air cavities. Light weight woods have thin walls, and large air spaces. The relative fuel values presume that you buy wood by the cord, a *volume* unit. If bought by the *pound* the fuel value of a heavy wood will be about the same as that of a light wood. This may not be apparent because heavy woods burn slowly, while light woods liberate their heat and are reduced quickly to ashes. The moisture in a piece of wood lowers its fuel value since considerable heat is used in raising the temperature of the water to 212°F, and still more in converting it to steam. Removal of water from wood also involves other heat losses.

Different kinds of wood vary greatly in the way they "talk" when burned. Some kinds burn quietly with hardly a sound, while others pop or explode, and throw out hot sparks or burning embers. Hemlock is a well known sparker, and according to Tiemann (12) such woods are relatively impervious to gases. Thus, internal pressures build up until the burning wood structure suddenly gives way. Considering the complexity of wood, there may be more to the story than this. In any case the firebuilder soon learns which woods spark and which do not.

All of the previous discussion about burning wood has assumed that the normal amount of oxygen in the air was present. When wood is heated in a closed retort excluding outside air

Figure 88 Redwood fire wall resists intense heat of rubber products fire. (Courtesy California Redwood Assoc.)

Figure 89 Redwood plywood box $4\frac{1}{2}'' \times 4\frac{1}{2}'' \times 10''$ inside. Three thicknesses of $\frac{3}{4}''$ plywood, recessed construction, resorcinol adhesive.

Figure 90 Redwood box exposed to fire for one hour at temperatures of 1600° to 1740°F Although the papers were safe, a crosswise fissure had developed in the bottom. Therefore, the thickness should be 3" all around.

(destructive distillation), a series of useful gaseous products can be piped off. Some of these upon condensation yield liquids. Heat must be applied until the inside temperature is about 525°F. At this point, the reaction becomes *exothermic* (gives off heat). According to Tiemann (12), the products given off in approximate order are water vapor and volatile oils; various gases such as carbon monoxide, carbon dioxide, methane (marsh gas), and hydrogen.

Except for the carbon dioxide, these gases can be burned as fuel at the plant. Then a dark liquid distills over leaving, finally, charcoal in the retort. After cooling and standing, the liquid yields a residue of tar, and a lighter colored water soluble liquid called pyroligneous acid. This in turn yields some oils and tar but, most importantly, acetic acid, wood alcohol (methanol) and acetone. These are three valuable chemicals from wood. A half dozen other chemicals in very small amounts may also be recovered. However, the hardwood distillation industry is all but finished. Its chemical products are synthesized in other ways more cheaply. Only charcoal continues to have a market.

When one uses the coniferous woods, the yield of wood alcohol and acetic acid is very low. The products of value here include oils, turpentine, tars, and charcoal. Among these softwoods, only pitchy heartwood of the southern pines, especially longleaf, is presently used.

It is easy to demonstrate the destructive distillation of wood. Place a few hardwood chips in the bottom of a test tube, and in the open end insert a cork with a vertical groove to serve as a gas vent. Holding the tube in a position somewhat above horizontal, slowly heat the wood end, and observe what happens. As the gases are driven off, they can be lighted at the vent in the cork, and the pyroligneous acid will condense as a dark yellowish liquid in the cooler, upper part of the tube. To explore this subject further see the book "Forest Products" (9).

Since ancient times, innumerable experiments have been performed to give wood greater resistance to fire. The early Egyptians soaked wood in an alum solution and the Romans did the same, with the addition of vinegar. Gay-Lussac, the famous French scientist, in 1821, tried many chemicals to flameproof cellulose fabrics. Recommended were mixed solutions of ammonium phosphate and ammonium chloride, and ammonium chloride and borax. Since then, hundreds of other chemicals have been tried, but these still remain among the best of fire retardants. Flaming of the combustible gases, and glowing of the charcoal finally remaining are two different processes, and each may require a different chemical for its control. In one series of tests, only ammonium phosphate and phosphoric acid retarded both flaming and glowing. A mixture of borax and boric acid may have a similar effect.

To return to the flaming gas stage at the start of a fire, the speed at which flames spread along the surface of the burning wood is of the greatest importance. Fire retarding treatments are intended to reduce spreading as well as postpone the first appearance of flame as the temperature rises. Mixtures of borax and boric acid in the outer layer of wood seem to form a glazed coating when heated. This coating slows down the emergence of inflammable gases, and prevents the entrance of oxygen. Even better are substances that upon heating form a barrier of heat resisting foam on the surface of the wood. Paints containing these chemicals can be purchased.

This entire subject is of enormous proportions with many hundreds of research papers describing experimental treatments and the development of theories from them. Recommended is "Theories of the Combustion of Wood and its Control" (3), as well as other publications on the subject by the U.S. Forest Products Laboratory.

Reaction Wood—Compression

How do leaning trees straighten themselves? The partial answer to this seemingly simple question lies at the end of a long and mysterious journey the details of which would fill a good sized book. To begin, look carefully at the cross section of the spruce (Fig. 91). Notice that there is a core of about twenty rings, and that each ring varies but little in width as you follow it around in a complete circle. Below and above this core of normal rings something remarkable has happened to all the other rings up to the time the tree was cut. At the top of the section, each ring is abnormally narrow. Follow one of these rings and you will see how it widens out, and changes from light (earlywood) to dark. With this continuing growth of rings that are very narrow at the top and very wide below, the center of growth is no longer at the center of the section but considerably above it. In conifers, the dark often reddish colored wood is formed on the *under* side of a leaning tree, and is called *compression wood*. This tree began life growing straight up, but after about twenty years something knocked it over so that its trunk formed an angle of about 30° with the ground. Then began the formation of compression wood and the development through the years of a smooth curve upward in the tree's trunk near the ground. This is how relatively small trees straighten themselves. When a large tree is upset, something else may occur. Although compression wood begins to form all along the under side of the trunk, the vast resistance of so large a stem makes recovery very slow except at some place near the tree's top. Here, the force developed by the compression wood is enough to make the stem curve smoothly, elevating it to a

Figure 91 Red spruce. Cross section showing normal wood above, compression wood below (tree leaned toward bottom of page)

vertical position. As many more years pass, the heavier part of the trunk below the bend may slowly curve upward and this causes the vertical top to lean once more. To straighten *this* lean, compression wood forms on the *reverse* or *new under* side of the stem. If this double process continues, the tree trunk in time may show a double curvature.

Most kinds of conifers (except the low-growing forms) develop a straight central stem from the very first and maintain it at least until they are very old when the top may become branchy. A few exceptions include the hemlocks which develop drooping new terminal growth each year. For a few weeks, the growing tip may point to the ground, but by the end of the season, the drooping year's growth has straightened itself or nearly so. Final straightening occurs during the next one or two growing seasons. Sometimes at the end of the season, the new leader may still have a nearly 90° bend from the vertical. What do we find developing all along the under side of these curved leaders? *Compression wood.* It is this that during the first season may straighten the top of a leader by nearly 180°.

To see how different compression wood tracheids are from normal ones, study the two enlargements below. You can see that those of compression wood are relatively small, are rounded in outline, and that in the "corners" the true middle lamella may be lacking. Also, the S3 layer is missing. The seemingly radial checks extend almost through the S2 layer and stop just short of the S1. These are actually the cross sectional view of helical checks in the tracheid wall. In lengthwise section, the electron microscope shows that the microfibrils run at about 45° to the long axis of the tracheid. Compression wood has more lignin and less cellulose than normal wood. Developing on the under surface of a leaning trunk, compression wood must expand or push lengthwise to straighten it. When a piece of compression wood is cut from a tree, the wood does indeed expand slightly lengthwise showing that in the tree, it was under compression.

Tension Wood

A leaning broadleaved tree develops an entirely different kind of reaction wood. Instead of occurring on the *under* side as in the conifers,

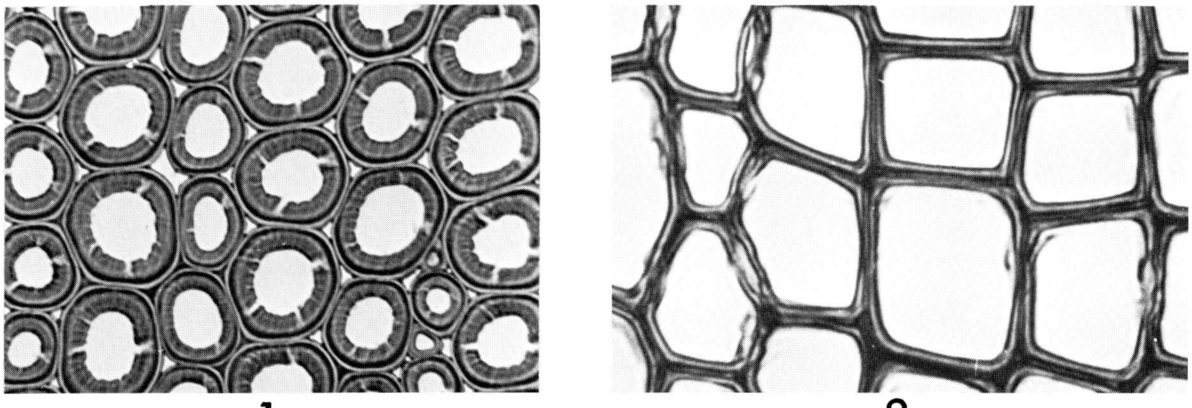

1. 2.

Figure 92 1. Compression wood tracheids × 750. (Courtesy W. A. Côté Jr. and A. C. Day) 2. Normal wood tracheids × 750.

it forms along the *upper* side. To straighten the tree, it must exert a pull and is therefore called *tension wood.* Why conifers develop compression wood and broadleaved trees tension wood is one of the unsolved mysteries of the plant world.

Now look at the cross section of the American beech (Fig. 93). The eccentricity of the ring pattern is so great that the center of growth is nearly at the bottom of the section. Actually there are two growth centers, showing that at the time the little tree was bent over it had two stems which soon grew together into a single trunk. The dark-colored wood below is normal, the large V-shaped area above is tension wood. The section was cut, near the ground, in the curved part of the tree. Although the tree had long since straightened itself, tension wood kept forming in the horizontal part of the curved trunk. The photograph was taken of a rough-sawed section since for some unknown reason, sanded and polished sections of such trees do not show well the color differences between tension wood and normal wood.

Typical tension wood cells show a so-called "gelatinous" or G layer inside the S2 layer (S3 is usually lacking, but may be present; S2 may also be missing). Under the microscope, the G layer looks somewhat puffy, and some early botanist may have thought it looked "jelly-like", hence the name. Modern research indicates that such a name is very misleading. Actually the G layer is principally cellulose of a highly crystalline form organized in microfibrils that run almost parallel to the fiber's long axis. The volume of the G layer that often nearly fills the cell cavity, and its cellulose structure should make it especially strong in tension. But its connection with the rest of the cell wall seems to be very poor. How can such a weakly attached layer furnish the tensile strength necessary? This is a mystery which deepens when we find that the tension wood cells in many kinds of hardwoods have no G layer at all. Externally, tension wood boards are often covered with bundles of fibers torn loose by the saw. The surface of such a board looks and feels "woolly".

Reaction Wood Problems

Both compression wood and tension wood are very definitely defects in the manufacture of lumber and other wood products. Normal wood during drying usually shrinks lengthwise not more than .1 to .3 of one percent. Compression wood shrinkage is very variable but it is often *ten times* or more as great as that of normal wood. When a drying board contains both normal and compression wood, the difference in relative lengthwise shrinkage often causes the board to twist and cup to a point where it is useless. Lengthwise shrinkage in tension wood rarely exceeds one percent, but the woolly surfaced boards and other effects make tension wood undesirable. Boards containing it also twist out of shape in drying.

Upon these few pages, it is hoped that enough has been written about reaction wood to interest you in exploring the subject further. A. H. Westing (17) in two recent reviews on the formation and function of compression wood consulted some 700 references or scientific papers, and yet many unanswered questions remain. You may have wondered just how the forces of "push" or "pull" are developed. There seems to be no answer as yet. We must find out what happens in the cambial cells, how they produce reaction wood, and how this exerts its forces. When one tries to study cell physiology and chemistry the enormity of the problems involved seems almost insurmountable. The biochemist Hofmeister made a rough calculation that in a single typical liver cell there might be about 53 billion protein molecules, 166 billion fat molecules, 225 trillion water molecules, and nearly 3 trillion molecules of miscellaneous substances. In any living cell including those in the cambium, the billions of molecules are constantly in motion acting and reacting with each other to form hundreds of different substances. The instant one probes into a cell, all these fantastic life processes are either altered or stopped altogether. More sophisticated techniques are being developed for exploring what goes on in living cells, and in time, hopefully, we may have answers to how reaction wood develops its tree straightening forces.

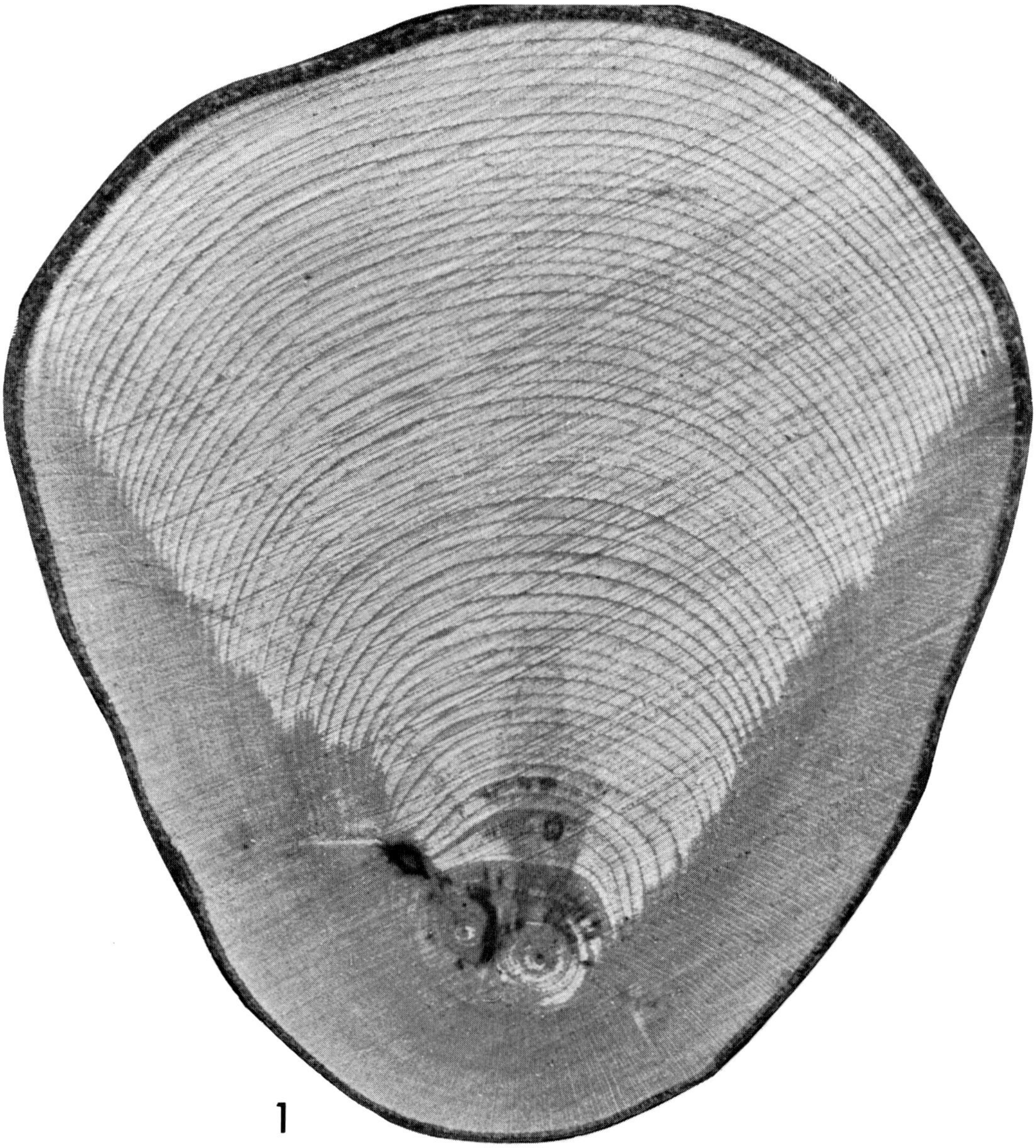

1

2

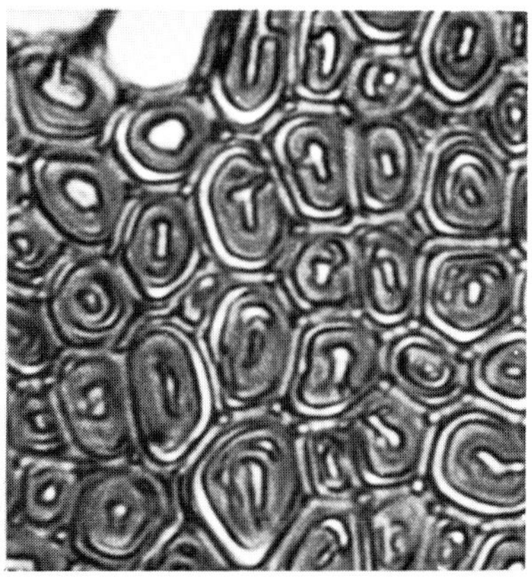

3

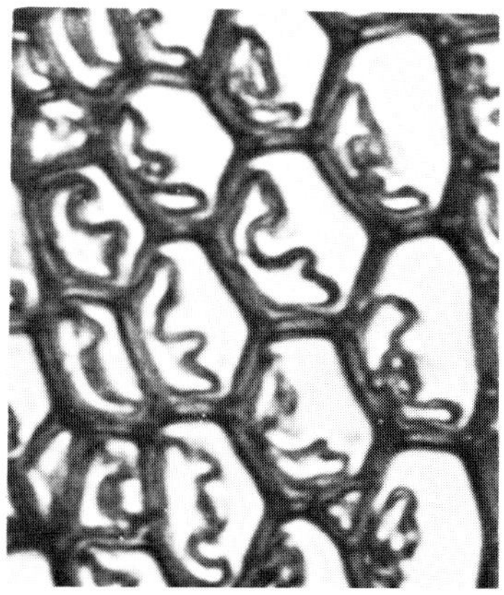

4

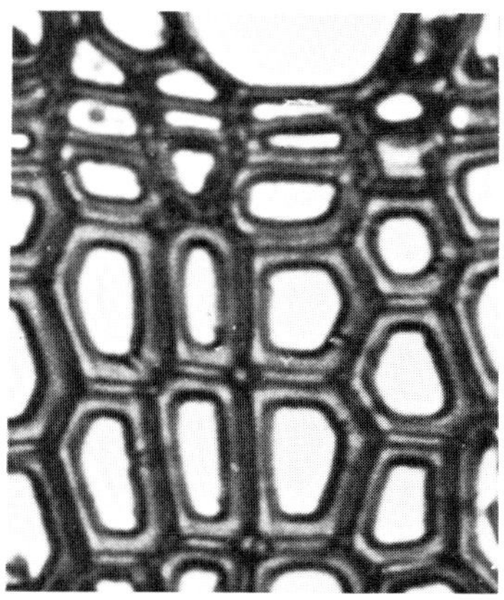

Anyone who walks through the woods can see enough crooked or curved tree trunks (especially in the hardwoods) to suggest that the tree straightening process is not perfect. Reaction wood may not always form, or is of insufficient quantity to complete the change of a leaning trunk to a vertical one. One other factor might be mentioned, inheritance (genetics). In certain species such as Scotch pine, there are races each of which has its own growth form. Some of these grow crooked "naturally", while others grow straight.

The spruce section Fig. 94 shows the most remarkable growth pattern illustrated in the entire book. At first glance it appears perhaps that the tree grew in a spiral which obviously is impossible. Enlargement of some of the rings shows quite clearly that seemingly normal wood (lower center) changes to compression wood as the rings curve upward to the left. From what we have said about compression wood, one would say that the normally straight tree suddenly leaned toward the upper left for three years, and then became straight again. This might be interpreted, but complete puzzlement occurs when one tries to explain the spiral development of compression wood above which spirals around the tree for the rest of its life.

The ring count is about 135, some of the rings being so narrow that it is difficult to see them even with a hand lens. This specimen was illustrated and written up many years ago in Ripley's "Believe-it-or-not" column, and the cross section presented to the N.Y. State College of Forestry. We do not know at what height in the tree the section was taken. If at some distance from the ground, then the tree was much older than 135 years. One thing we can say about it is that it grew its entire life in the forest and not in the open; the rings are very narrow. Many knowledgeable people have been astonished by this section, and have tried to explain it without success. You already know that in the conifers, compression wood forms on the under side of a leaning tree. Does this suggest that all during its lifetime, the tree's leaning trunk slowly moved in a circle? What would cause it to do this? Another idea is that the tree in reaching for the light developed a heavy crown on one side and that this development "moved" in a complete circle some three times during the life of the tree. What would cause this? Do you have any suggestions? Could it be an inherited trait (genetics)? Such a pattern is very rare and has been reported only a few times in widely separated places in the world.

The Strength of Wood

This is another vast subject, and its understanding should presuppose a knowledge of mechanics and strength of materials. However, as with the chemistry of wood, an attempt is made in this case to highlight for the beginner the behavior of wood in response to forces acting upon it. First of all, for its weight, wood is an immensely strong material. Look at Figure 95. Here is a small block of Douglas-fir wood one inch square on the ends, and two and one quarter inches long. In a testing machine, it was loaded slowly with increasing force parallel to the grain until it failed as shown. Just before it did so, this little block weighing only about three-fourths of an ounce supported a weight of 10,000 pounds, or the combined weight of three automobiles.

A few basic terms in mechanics must now be introduced. The first of these is *force*. We have all observed the action of wind forces, the force of moving water, and that of a jet engine. All of these can move something, or if the something is

Figure 93 **American Beech. 1. Cross section showing normal wood below, tension wood above (tree leaned toward bottom of page) 2. Tension wood fibers × 750. 3. Aspen tension wood fibers × 750; sectioning knife tore away the loosely attached "G" layer. 4. Normal fibers × 750.**

Figure 94 Spruce cross section showing spiral pattern of compression wood.

Figure 95 Small block of Douglas-fir wood broken by compression parallel to the grain.

already moving, cause it to go faster, or slower, or change its direction, or stop, depending upon the amount and direction of the force applied.

There are three principal kinds of force that can be applied to a piece of wood or other material: (1) The "pushing" or *compressive* force applied to the small block of Douglas-fir wood where, as more and more force is applied, the block gets shorter. (2) A "pulling" force applied in the opposite direction to No. 1. Here, the test piece is stretched and becomes longer; this is *tension.* (3) Reference to Fig. 96 will make clear what happens when a test piece is subjected to a *shearing* force. The piece fails along the dash line by one half of it slipping past the other half.

These three forces may be called contact forces. All bodies and structures are also acted upon by the force of gravity, one of the unsolved puzzles of the universe. The attracting force of the earth's gravity on a body is called its *weight* which may be measured in pounds. Compared to the enormous compressive force required to break the Douglas-fir block, the force of gravity in this test is insignificant since the block weighs only about three-quarters of an ounce. However, in a long wooden beam supported only at the ends, the weight of the beam itself may be considerable and must be accounted for in figuring the additional load that the beam will carry. The distribution of forces in a beam is quite complicated. Fig. 97 shows a beam broken by a force of 3,000 pounds applied in the center of the upper surface. As the bend developed, the wood along the bottom of the beam was under more and more *tension,* while that on top was under increasing *compression.* Finally the beam broke, mostly by the wood being pulled apart (tension failure). It is not easy to see the results of compression failure on top but they are there nonetheless. In a beam of uniform composition (homogeneous), it might be expected that through the center lengthwise there would be a horizontal *neutral plane.* Here, there would be neither compression nor tension, but rather shear. Moving toward the top, compression would increase to a maximum at the upper surface. Moving from the neutral plane downward, tension would increase to a maximum on the lower surface. Wood, with its great complexity of structure rarely follows an exact pattern, and it is clear that in this beam (Fig. 97) most of the failure was by tension which developed well above any theoretical neutral plane.

To return to the idea of force, a force applied to a unit area or volume is called a *stress.* In the case of the Douglas-fir block one inch square on the end, the 10,000 pounds would be the stress (10,000 pounds per square inch). If a block two inches square were tested, its end area would be four square inches, and the stress would be the number of pounds divided by four. *Strain* is a measure of the unit deformation of the test sample. One would measure the amount of shortening (deformation) of the loaded test piece and divide this figure by 2¼ inches, the length of the Douglas-fir piece. Strain figures are extremely small for wood, and the readings are in thousandths, or even ten thousandths of an inch. Unfortunately, in general usage, the words "stress" and "strain" have various meanings and at times may be used as synonyms, or even given reverse meanings. Their use in physics is an example of how in science common words of

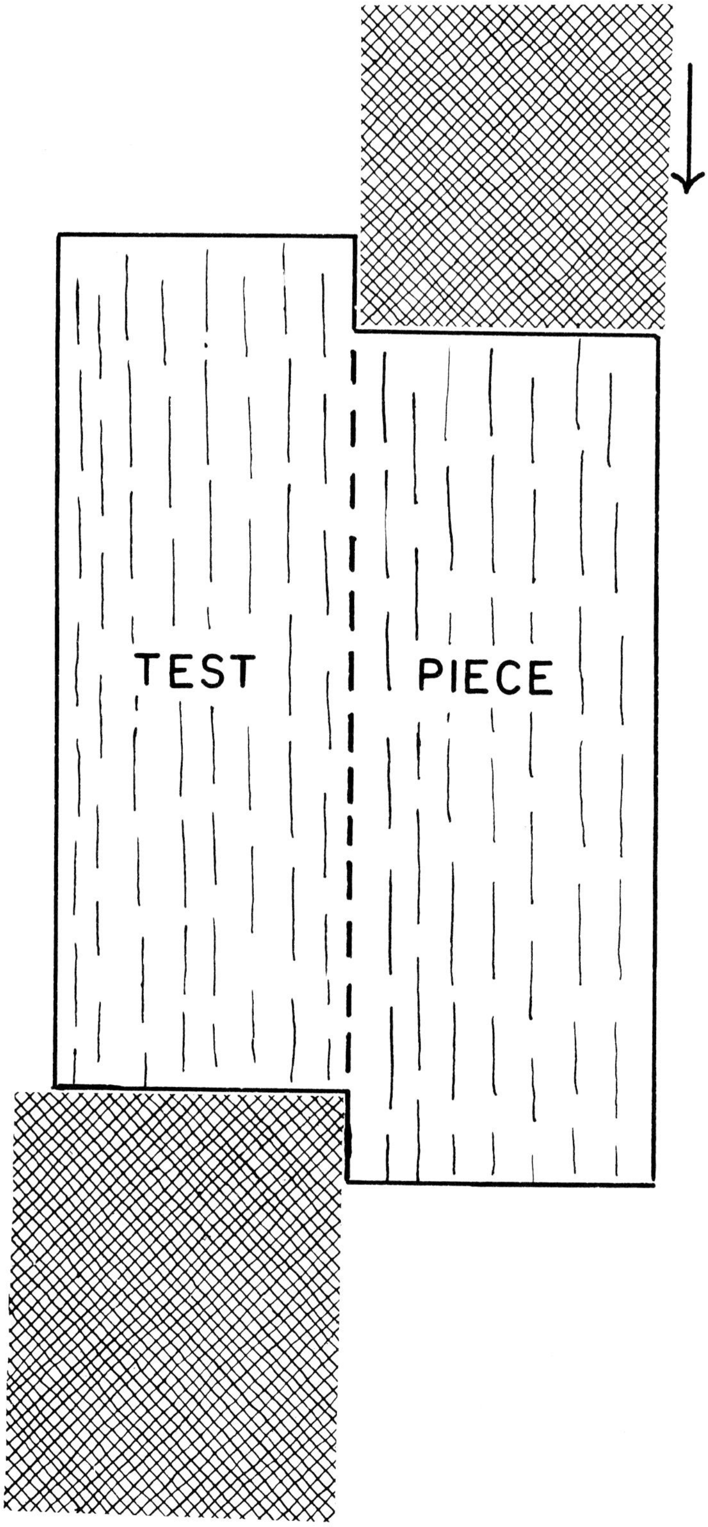

Figure 96 This diagram shows how a piece of wood is tested to discover its resistance to a shearing force.

unclear meaning may be chosen and given a precise definition.

Stress divided by strain $\frac{\text{stress}}{\text{strain}}$ is a measure of the *elasticity* of a piece of wood or other material. To most people a rubber band is highly elastic, but in physics it is just the reverse; rubber has a very low measure of elasticity. Technically, elasticity is the ability of a material to recover exactly its original form when released from a stress. Perhaps for the general reader the term stiffness might have more meaning. If you take a thin strip of wood, hand held at each end, you can bend it. If it bends easily, it is not very stiff (elastic); if it is difficult to bend, then it has considerable stiffness (elasticity). When testing a material with a uniformly increasing load, stress and strain maintain a more or less constant relation to each other until the *elastic limit* is reached. At this point, if the stress is removed, the piece of material will return to its original

Figure 97 Beam of white ash broken by loading on upper surface, center.

shape and size. In doing so it will perform work. Consider a flexed bow and the flight of an arrow as the bow string is released. Beyond the elastic limit, as the load continues to increase, the relation between stress and strain is no longer proportional, and the process of failure or breakage begins. If complete breakage occurs almost at once, the material is considered to be *brash* or brittle. If it yields slowly and breaks gradually it is said to be tough.

Early testing was conducted mostly with metals, and concepts and procedures were developed for them. Wood with its almost infinitely greater complexity of structure often does not follow precisely the picture just sketched. However, using small test pieces, stresses building to a maximum within minutes or less, and strains of small amount, wood behaves essentially as described.

The resistance of a piece of wood to forces deforming it is called its *strength*. There are several factors that affect strength. Basically, specific gravity, a measure of the amount of solid cell wall material in a piece of wood is the best indicator of its strength. In compression parallel to the grain, extractives if present in large amounts often increase the strength. In wood, the structuring of the material into walls of hollow tubes gives wood a flexural rigidity exceeding that of solid metal structural members of the same weight. This structure of wood, however, makes it weak in shear, and even not as strong in compression as some of the metals. A good method of measuring tension along the grain in a strip of wood has yet to be found. If one bores a hole near each end, inserts bolts and then applies a pulling stress to the bolts they will pull out of the wood when a relatively small force is used. The wood has failed by shear, not tension. Also the piece has been weakened by the holes themselves. When end clamps are used to hold a test piece, then the wood structure crushes in the clamps. It is much weaker across the grain than it is lengthwise. As we have seen, tension breaks occur on the under side of a stressed beam. When a tree is felled, there is usually a section of undamaged wood left between the two opposite cuts that have been made. As the tree falls, the wood in this section usually fails in tension. An excellent example is shown in Fig. 98. Wood is considered to be stronger in tension than it is under compression.

Most of the strength properties of wood are highest when it is ovendry. As moisture is added, strength decreases until it is lowest at the fiber saturation point. Ovendry wood may be from two to four times stronger than green wood. Also the type of break is different. When a short piece of dry wood fails under compression parallel to grain (Fig. 95), the edges of the break are usually sharp and distinct. By compression, the fibers in a green piece bend and buckle giving way slowly and not splitting apart suddenly as they may do when dry. The dry piece is brash or brittle compared to the wet one. From what was said above about brittleness and toughness, it seems clear that as moisture is added to dry wood it becomes "tougher". This reverses the general rule about dry wood being stronger than wet wood. Toughness is the ability to keep on absorbing force past the elastic limit and failing slowly as it does so. The shock resistance of wood is also affected in a similar way. A dry ax handle (helve) is more brittle than one that has some moisture in it. The following table with figures from the "Wood Handbook" (13), illustrates the effect of both moisture and specific gravity on two strength properties of wood.

These figures are averages derived from a number of tests using small straight-grained pieces without knots or other visible defects.

Temperature is another factor affecting the strength of wood, but in most liveable regions of the world the range in temperature is not enough to cause more than small changes in strength. At 100°F the compressive strength parallel to the grain of a piece of fir wood was found to be 11,000 pounds per square inch, while at 50°F below zero, the strength had increased to 13,000 pounds. Using an extreme temperature range from 300°F down to 300°F below zero, the hot piece of wood supported 7,900 pounds while the cold piece resisted a force of 17,000 pounds. Wood samples heated to no more than 150°F for short periods of time and then returned to room temperature show no loss in strength prop-

Figure 98 Tension break on a stump of black cherry × 6.

The Effect of Moisture Content and Specific Gravity on the Strength of Wood*

(All values in pounds per square inch)

Species	Percent Moisture	Specific Gravity	Compression parallel to grain. Crushing strength	Shear parallel to the grain
SOFTWOODS				
Western larch	58	.51	3,990	900
	12	.55	8,110	1,410
Eastern white pine	73	.34	2,440	680
	12	.35	4,800	900
HARDWOODS				
Shagbark hickory	60	.64	4,580	1,520
	12	.72	9,210	2,430
Yellow-poplar	83	.40	2,660	790
	12	.42	5,540	1,190
Quaking aspen	94	.35	2,140	660
	12	.38	4,250	850

***These figures are from the "WOOD HANDBOOK" (13).**

erties. Those heated above 150°F for some time and then tested at room temperature are weaker than they were before, and this weakness increases as higher and higher temperatures are used up to the point of ignition. The weakening is caused by chemical changes produced by the high temperatures and other factors.

Another characteristic of wood is its plasticity. That is, it tends to "flow" or *creep* if continuously stressed. This is especially true in bending, and accounts for the sagging of beams in old structures that have carried loads for long periods of time. However, even though there is a loss in strength, the size and number of timbers used in building construction keeps these structures safe no matter how long they may stand.

If you fasten one end of a strip of wood in a vise and then bend the strip back and forth continuously, another feature of wood is displayed, —its resistance to *fatigue*. This is extremely high compared to some of the metals. Using a stress of about one third that required to break the wood, it can be bent back and forth some 30 million times before failing.

There are still other things besides specific gravity, moisture content, and temperature that affect the strength of wood. These include crooked grain, knots, reaction wood, the width of the growth rings, the relative amounts of earlywood and latewood, and the part of the tree (near the center, or farther out) from which the piece was cut.

To anyone with an engineering background, the adapting of wood for new uses based on its unique properties is indeed a challenge.

Chapter VI

Decay by Fungi

The question here is how best to begin this story of such great importance to users of wood throughout the centuries. Most of you have probably seen and touched decayed wood. You know that it is weak and easily broken, often spongy, that it holds water, shrinks more than sound wood when dried, and is usually differently colored. When walking in the woods or city parks, you may have seen one or more of the often highly colored, beautifully structured fruiting bodies of various fungi growing on live tree trunks or on dead trunks and branches lying on the ground. Where old, wooden buildings are being torn down, you may have seen these fruiting bodies growing from a flat whitish mass that grows over the surface of the wood. This mass is the vegetative part of the fungus and usually develops like this only where the humidity is high such as in enclosed spaces with poor ventilation. Fig. 99 shows a fungal mass *(mycelium)* growing on a small block of ponderosa pine, kept in a warm moist chamber for one year.

A good place to start developing the relation between fungi and wood decay is at the fruiting body. The one shown in Fig. 101 of the rainbow fungus is so-called because of the concentric "rings" of different colors. Looking at the under surface, you can see with a hand lens that there are hundreds of tiny pores. From these openings stream forth not millions but billions of microscopic spores each one about one ten-thousandth of an inch in diameter. These are carried everywhere by wind currents. When a spore lands in a favorable spot, one or more tubes with rounded closed end walls begin to grow outward from the spore. Each tube contains protoplasm and is called a *hypha* (pl. *hyphae*). From the tips of the hyphae which branch and rebranch as they grow to form the mycelium, there are secreted enzymes which initiate a whole series of chemical reactions which tear down the holocellulose and/or the lignin in the wood. The end products are carbon dioxide and water, but the intermediate products are passed in through the walls of the hyphae, and at each step in degradation, a simpler product is formed and some *energy* released which the hyphae use to live and extend themselves throughout the wood structure. The picture shows several hyphae growing lengthwise in the tracheids of a conifer, and one place where the growth is crosswise from one tracheid to the next. Here, as the tip touched the wall, the wood was dissolved, and the hypha passed through the hole. Unfortunately, the picture shows only fragments of the interconnected system of hyphal tubes. This is because the tubes grow in all three directions, and the section, cut about one thousandth of an inch thick, can show very little in the third dimension.

The processes of decay may be thought of as similar to those of photosynthesis in reverse. In the green leaf, water from the ground and carbon dioxide from the air, in the presence of chlorophyll, are built up through a series of beautifully complex chemical reactions to form simple sugars in solution. At each step, the chlorophyll captures some of the sun's energy and locks it up so to speak in the substances synthesized. The sugar solutions circulate in the tree's phloem (inner bark), and in the cambial cells are further fabricated into the more complex molecules of holocelluloses and lignin. All of the stored energy comes from the sun through photosynthesis. When a fungus uses wood substance as food, it absorbs this energy to carry on its own life processes and extend its body throughout the wood.

After decaying the wood for from one to many years, certain hyphae near the surface of a tree, or piece of wood begin to form the fruiting bodies (sporophores) which emerge and then

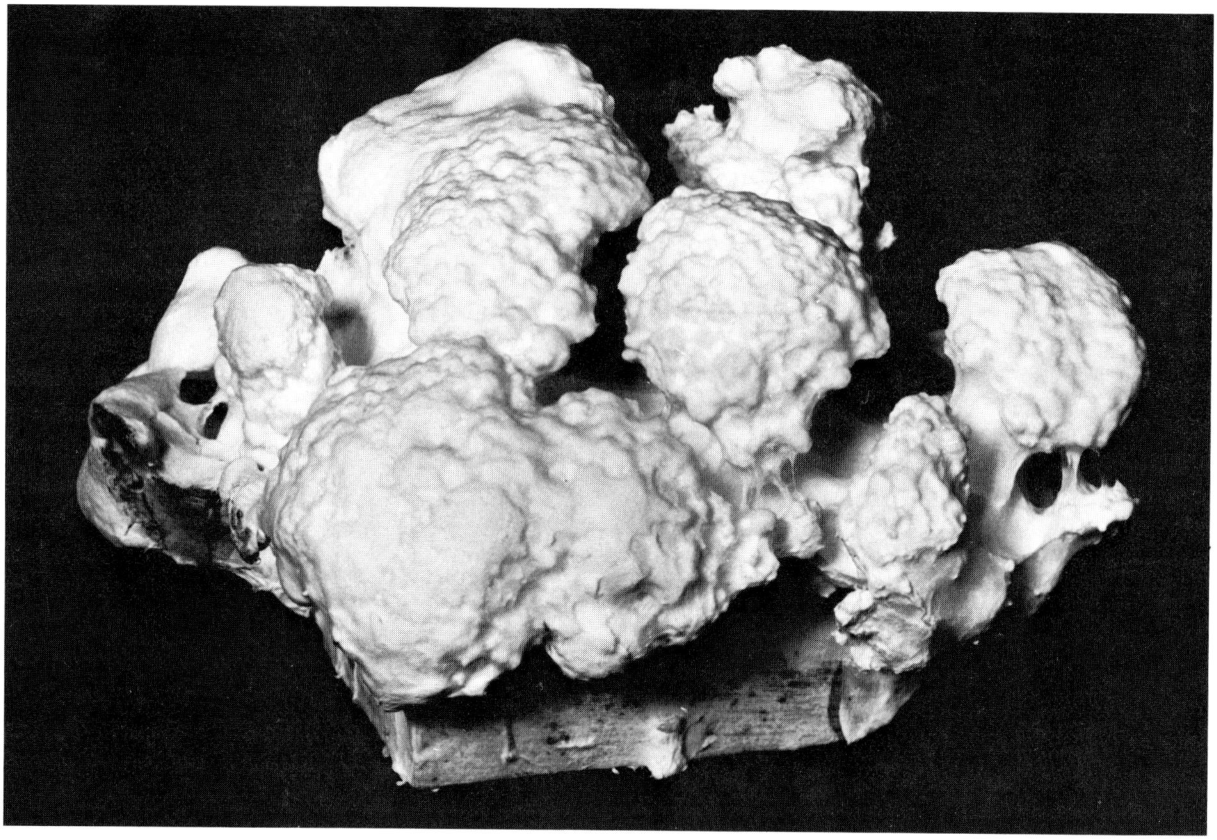

Figure 99 One year's growth of a wood-destroying fungus on a block of ponderosa pine wood kept in a warm moist chamber.

Figure 100 After one year, this block of oak lying beside the pine block still resists the fungus.

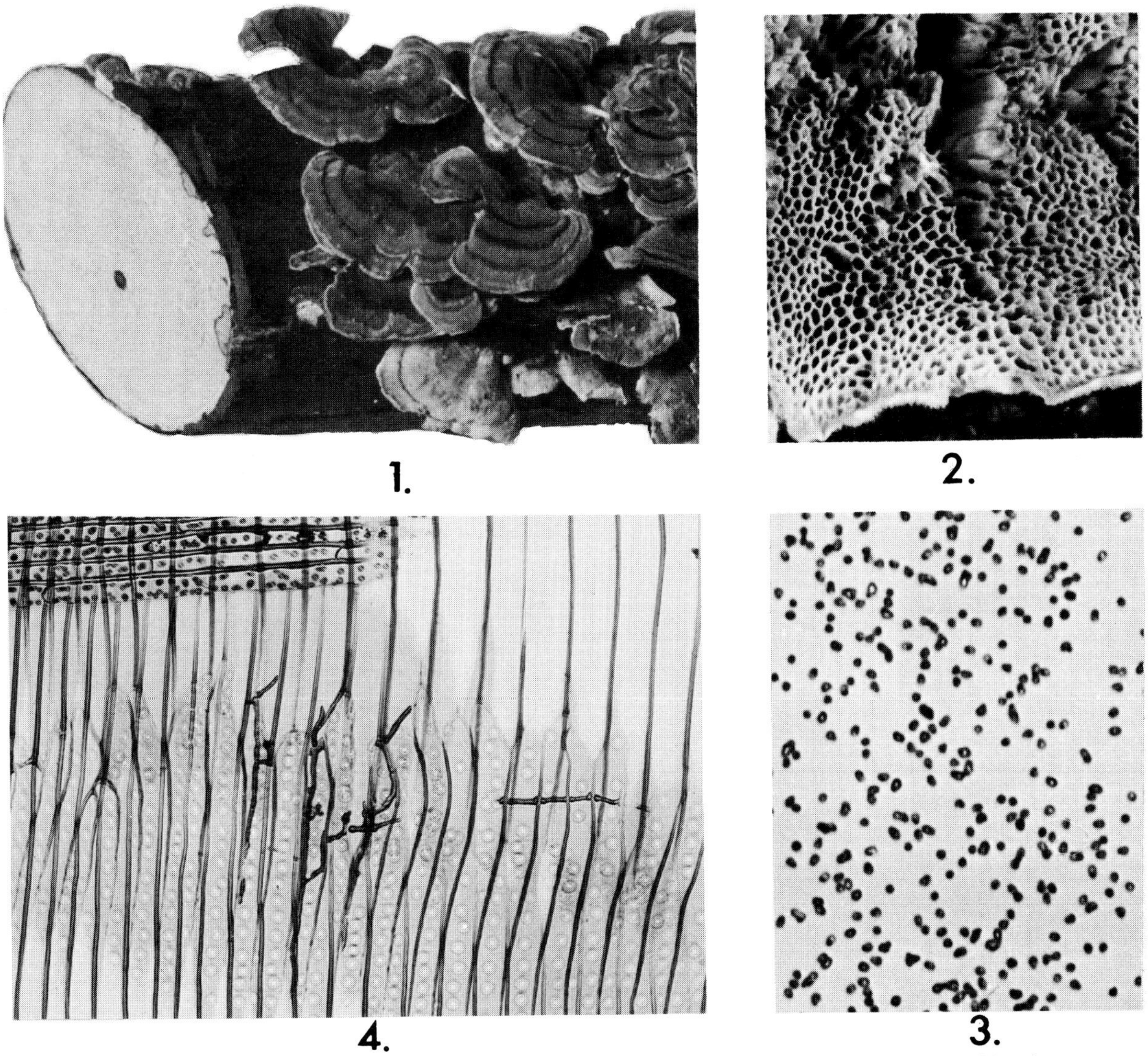

Figure 101 **1. Rainbow fungus; white rotted wood, and fruiting bodies. 2. Under side of fruiting body shows spore exit pores × 8. 3. spores × 1,000 4. hyphae growing in a coniferous wood × 250.**

may continue to grow for some time often with yearly additions. It should be made clear that the sporophore is the final stage in the life history of a fungus, and that when one or more appear it means that the tree or timber is already decayed internally. Some people have thought that by cutting off the sporophores, they could stop the decay. This certainly is not so.

Conditions necessary for decay to develop are very important to know. First of all, fungi must have moisture, Keep wood dry and it will never decay—but how dry? There are so many kinds of fungous wood destroyers, and so many kinds of wood with varying resistance to these fungi that general statements can only be approximate. Moisture contents of 35 to 50 percent are favorable for fungal growth. Between these levels, the cell walls are saturated, and a thin layer of free water coats the inside of the cell cavities. To be safe from attack, wood must have a moisture content somewhat below that of the fiber saturation point. For most temperate climate woods, the f.s.p. is about 25 to 30 percent, and fungal growth ceases at 20 percent moisture content or less.

Most home owners do not understand the importance of keeping wood as dry as possible. Wooden house or porch members must *never* touch the ground. Any wooden parts should be at least 8 inches from the ground, and porches should be well ventilated underneath. It must be said that architects and builders often neglect measures that would reduce or postpone decay. Every piece of construction should be designed so that rain water runs off and does not stand in puddles, or have a chance to penetrate joints where dampness may remain. Another danger may exist in insulated, heated houses in winter. Warm air from inside passes through the plaster or sheetrock inner wall and condenses on the cold outer wooden walls. Unless during construction they have been protected by a vapor barrier such as asphalt paper or aluminum foil, this condensation may raise the moisture content enough to allow decay to start. In older houses, a good vapor barrier paint on the inside should prevent this from happening.

Besides moisture, fungi require a favorable temperature for best growth. This seems to be between 76 and 86 degrees F. For active growth 40 degrees is about the minimum, but they do however survive low winter temperatures. High temperatures kill fungi. In one experiment, using a number of kinds of fungi, it was found that they were killed in 12 hours by 131 degrees of moist heat. When dry heat was used they did not die until the 221 degree mark was reached.

In addition to favorable moisture and temperature levels, fungi require oxygen for respiration. In only a few situations can they be robbed of this essential. Water-logging is the commonest. If a log or timber in the green condition has enough water in it so it will sink, the remaining air in the cells will be slowly replaced by water. Such logs or timbers have been "mined" from the bottom of a lake or river one hundred or more years later and found to be perfectly sound. Similarly, a constant water spray may keep a pile of logs wet and prevent fungi from entering, but will not kill those that may be present already inside. A common example involving both moisture and oxygen is what happens to a cedar wood fence post set in the ground. After a number of years, it will show decay just about at, and just below the ground level. Obviously, moisture conditions at these levels are favorable to wood destroying fungi. Above, the wood stays too dry, and below in the moist or wet earth, the oxygen supply is not enough for decay to occur excessively in the buried end of the post.

When it comes to the fourth and last essential, food, the situation is of the greatest imaginable variation and complexity. It is perhaps not possible to make general statements that will cover all cases. First of all, heartwood, especially that in which chemicals such as phenols and tannins are found, resists decay long after the sapwood has been destroyed (Fig. 102). Fungi may be grouped as sapwood destroyers or heartwood destroyers. Some do both.

So far, no distinction has been made between fungal decay in living trees, and that in wood products. A living tree with its protective bark is a marvelous organism that has defenses against invading fungi even after the bark is ruptured by a broken branch, or torn away by a passing logging machine. Damaged but still live cells in the exposed wood may produce chemicals that resist the invasion of micro-organisms. In spite of this, the spores of some wood-destroying fungi require such a freshly exposed surface to start growth. Many other kinds, however, cannot enter until several to many years have passed, and the wound has been prepared. First, the exposed white sapwood is darkened by exposure to air. This discoloration may penetrate into the tree as

Figure 102 Cross section of fallen dead oak tree lying on the ground. The sapwood is heavily decayed. The heartwood is resistant to decay. There are several different fungi decaying the sapwood; each one gives a different color.

well as being on the surface. Then, bacteria invade along with certain fungi. These fungi do not cause decay but are associated with the discoloration process that often spreads to the center of the tree and extends a long distance above and below the level of the broken branch stub or other wound. The whole heart of the tree may be dark reddish brown. After these events have occurred, the wood destroyers enter. However, there is a considerable overlap in the development of these stages. They are not sharply separated in time. A Photo Guide to the patterns of discoloration and decay in living Northern Hardwood Trees (10) carries this discussion much further, and mentions a review with 300 references on successions of organisms in the decay process. Decay of dead trees, and forest products is more "direct" since the barriers that the live tree interposes are no longer being built. Even so, there is still competition between organisms for the food materials.

There are three principal kinds of decay found in wood:

White Rots are found mostly in the wood of the broadleaved trees (hardwoods). Fig. 101 showing the fruiting bodies also shows the white surface of the cut end of the decayed branch. When dry, this branch feels fairly solid although very light in weight. When soaking wet, it is soggy and easily crushed by hand. In the white rots, the fungus digests both the holocelluloses and lignin, but mostly the former. Final reduction of white-rotted wood to carbon dioxide and water is done by other kinds of fungi and bacteria. The process is helped along by the wood-eating, tunneling larvae of certain insects, and by other micro-animals such as nematodes which are microscopic eels.

Brown Rots occur principally in coniferous woods (softwoods), and besides the color difference, there is a much different structural pattern. The decayed wood tends to fall apart into small blocks (Fig. 105, No. 2). This pattern is quite similar to the blocky appearance of a burned wood surface, or charcoal blocks themselves. Why such a pattern should develop in brown rots, burning wood, or either one, appears to be a mystery. Nothing in the ultra-structure of woody cell walls seems to explain it. Someone has coined the phrase "the slow fire of decay". This is something to think about since the final products of both wood decay and combustion by fire are carbon dioxide and water. In the brown rots, it is the holocelluloses that are attacked, and digested, but the lignin is also degraded even though not much used by the fungus. The final stages of decay are accomplished by other organisms as in the white rots.

Soft Rots are of less importance than the two previous ones, and are caused by a different group of fungi. Soft rot is found under conditions of excessive moisture where the wood is wet most or all of the time. It can usually be scraped off and underneath will be found relatively sound wood.

Dry Rot is a misleading term since dry wood does not decay. This is one of the brown rots and is caused by two fungi, one originally native to Europe, the other to North America. Dry rot fungi invade the basements of buildings, along the waterline in the holds of wooden ships, and other such places where moist wood is found and ventilation is poor. After the fungus is established, its body (mycelium) begins to spread in fan-like sheets moving outward toward parts of the wooden structure that are too dry to support fungal growth. But the dry rot mycelium contains special vein-like structures that carry water along with it, which raises the moisture content of the dry wood ahead to a point where the hyphae keep growing. In this way dry wood is often badly decayed.

Through the ages, wooden ships have suffered ceaseless damage from the silent creeping growth of the dry rot fungi. Dry rot destroyed more wooden ships than all the naval wars of history. The mysterious disappearance of old sailing ships at sea has been laid to the fact that dry rot had destroyed the wood all along the waterline. Then in a gale, the entire bottom suddenly dropped out and the ship sank.

A British admiral whose flagship was leaking badly decided to make a careful inspection. "We began by discovering slight defects in the ship, and the farther we went in the examination, the more important they appeared, until at last it

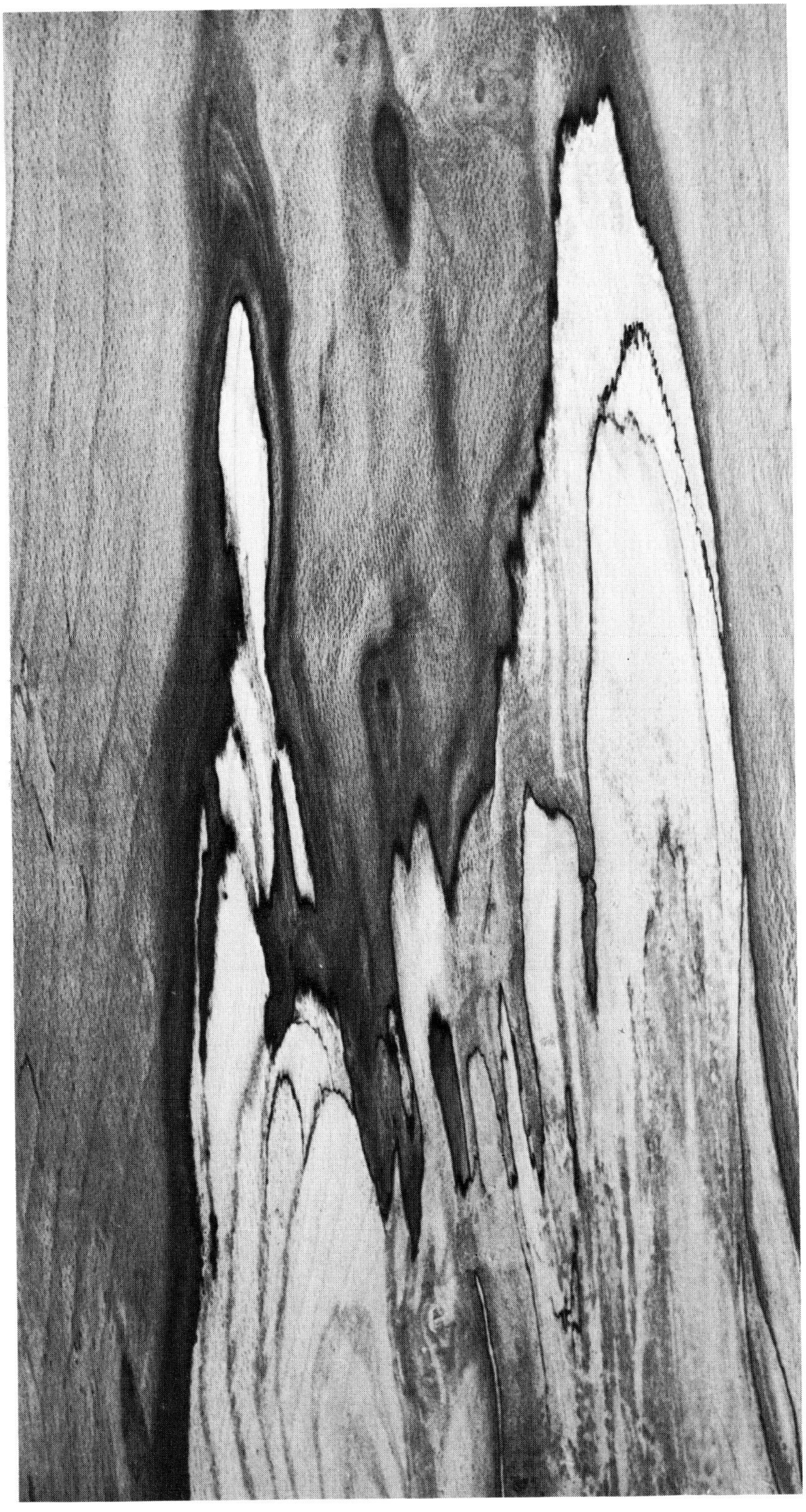

Figure 103 Beech showing invasion of wood destroying fungus causing white rot.

Figure 104 This birch cross section is a classic example of what often happens when a tree is blazed, or the bark knocked off exposing the sapwood. See how each new successive growth layer from both sides seems to "flow" over the face of the wound. Although finally healed, the decay that started on the exposed face has spread extensively. The dark discolored areas around the decay proper are presumably caused by other fungi than those that caused the decay (see text p. 102)

1. 2.

Figure 105 1. Fungal body (mycelium) of a wood-destroying fungus. 2. Brown rot decay in pine.

was discovered to be so completely rotten as to be unfit for sea. We have been sailing for the last six months with only a copper sheet between us and eternity" (2).

Sap Stain is often seen on the surface of sapwood in green, freshly cut boards, during humid weather, at temperatures of 75 to 85 degrees F. Various molds of several colors may grow vigorously on lumber piled for drying. These may cause surface stains. The true sap stain fungi go deeper and produce colors not easily removed. This may cause the wood to be rejected for certain uses where such staining is considered a defect. Mostly, the sap stain fungi are not wood destroyers, but find their food in the cell contents of wood ray and other live cells. Certain of them may cause a slight degradation of the cell wall. Besides the unwanted variously colored stains produced, it should be remembered that the moisture and temperature conditions under which these fungi grow are the same as those needed by the wood destroyers.

Wood decay causes hundreds of millions of dollars damage and loss each year both in standing trees, manufactured wood products, and wooden structures of all kinds. To prevent, stop, or slow down the decay of wood products is the purpose of the wood preservation industry. It is beyond the scope of this book to explore the methods of wood preservation. Poisonous chemicals such as creosote, copper naphthenate, and pentachlorophenol are forced into the wood by pressure treatment in large cylinders. Such treatment increases for many years the useful life of railroad ties, timbers and lumber that is used where the risk of decay is high.

Figure 106 Laminated wood beam construction. Photo by American Institute of Timber Construction.

Chapter VII

A Few Uses of Wood

On pages 40 to 53 there are listed some of the almost innumerable uses of wood. Many books could be written describing the features of the thousands of different woods in the world, and how these features make certain kinds so useful for general or specific purposes. Also included could be how man takes wood and modifies or fabricates it into valuable new products never before seen.

As an example of a special use wood, lignumvitae for the propeller shaft bearings in ships has had no equal. This wood has the peculiar property of being lubricated by salt water. The bearing tube is lined with narrow strips of the wood bevelled so that there is a V-shaped groove between each pair of strips. Presumably the natural oils in the wood contribute to the lubrication. Of course modern technology has tried to find a good synthetic substitute, but there are still thousands of ships with lignumvitae lined shaft bearings. Currently, new ships have metal babbitted oil lubricated bearings, but whether these will be entirely satisfactory, only time will tell.

Many new and exciting uses of wood were not possible until the discovery of modern adhesives to replace the older animal glues that would not resist moisture, and also had other disadvantages. Synthetic resin adhesives, some of them stronger than the wood itself have made possible the fabrication of immense beams, arches, and other structural members of shapes limited only by the imagination of the architect. Unlike plywood made from thin sheets of veneer with alternating grain direction, laminated structural members are made from boards with the grain running lengthwise. In a single beam or arch, there may be hundreds of boards glued together under pressure in a special form or jig. Woods most commonly used include Douglas-fir, and the southern hard pines.

Auditoriums, churches, and especially gigantic field houses for athletic events are being built with these great soaring arches secured crosswise by heavy roof decking. This simple construction with no vertical supports below gives an interior of great beauty and a feeling of spaciousness. The arches are usually not painted, and the color of the wood is pleasing to the eye. Such arches have been built with a clear span (building width) of 300 feet, and even greater spans can yet be produced.

Besides being beautiful, these structures are fire resistant, stand against high winds, and come through earthquakes surprising well. There is also no corrosion problem as with the metals. Laminated beams, arches, and structural members of other shapes are opening a new age in architecture featured by buildings of great beauty and utility.

Plywood

Plywood is not new. The Egyptian Saggara pyramid, circa 2750 B.C., contained a plywood coffin. There were six layers, and the woods were cypress, juniper, Sidder, and Lebanon cedar. Good glues were not known; the wood layers were fastened together with rivets of gold.

We have already seen that tangentially and

Figure 107 **Edge of 5-ply redwood plywood × 4½ (core piece is Douglas-fir).**

radially, wood shrinks and swells. To overcome this, thin continuous sheets of wood are "unrolled" from a log that in a machine is pivoted at each end and turned against a large blade that is as long as the log. (Sheets of wood may also be made in a slicing machine.) As it comes from the log, the long continuous sheet of veneer is broken at intervals both for convenience and also so that matched grain surface patterns including those of sapwood-heartwood may be produced (Fig. 108). Plywood is made by alternating at 90° the grain direction of the plies as shown in the redwood plywood above. By doing this, swelling and shrinking in the plane of the fabricated plywood sheet is very little. Modern plywood depends a good deal upon the excellence of the chemical adhesives now used. The glue line is often stronger than the wood itself.

The resistance of plywood to moisture or immersion in water depends upon the kind of adhesive used. Ordinary plywood for indoor use will come apart if left outdoors in the rain. An intermediate "water resistant" grade is produced, and the ultimate is marine grade which is entirely waterproof and is used in boat and ship construction.

Besides the fact that plywood dimensions stay almost constant despite changes of moisture in the air, it has other advantages. Unlike boards, it can be fabricated into large standard sized panels. Occasionally, the grain, texture, and wood color in a single tree are unusually beautiful. Such trees are often sold at fantastic prices. Plywood with a face veneer of this figured wood allows many users to feel its beauty rather than just a few were the tree to be made into "solid" furniture.

Particleboard

During the last two decades, this interesting product made from wood waste or low grade logs, forest thinnings and pulpwood has established itself as a useful material especially in the building and furniture industries. In 1955, about 78 million square feet of particleboard were produced in the United States and by 1966 the

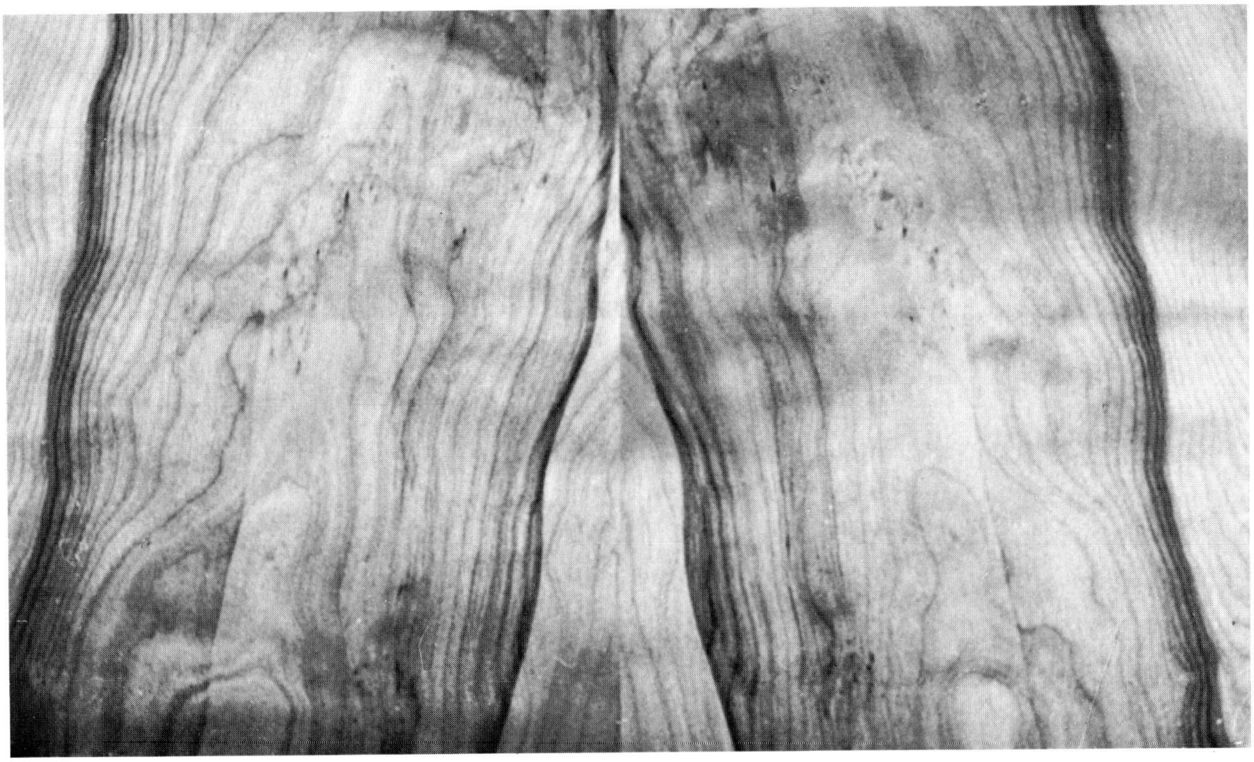

1.

Figure 108 No. 1 Surface matched veneer showing heartwood and sapwood pattern. No. 2. Rotary cut Douglas-fir plywood.

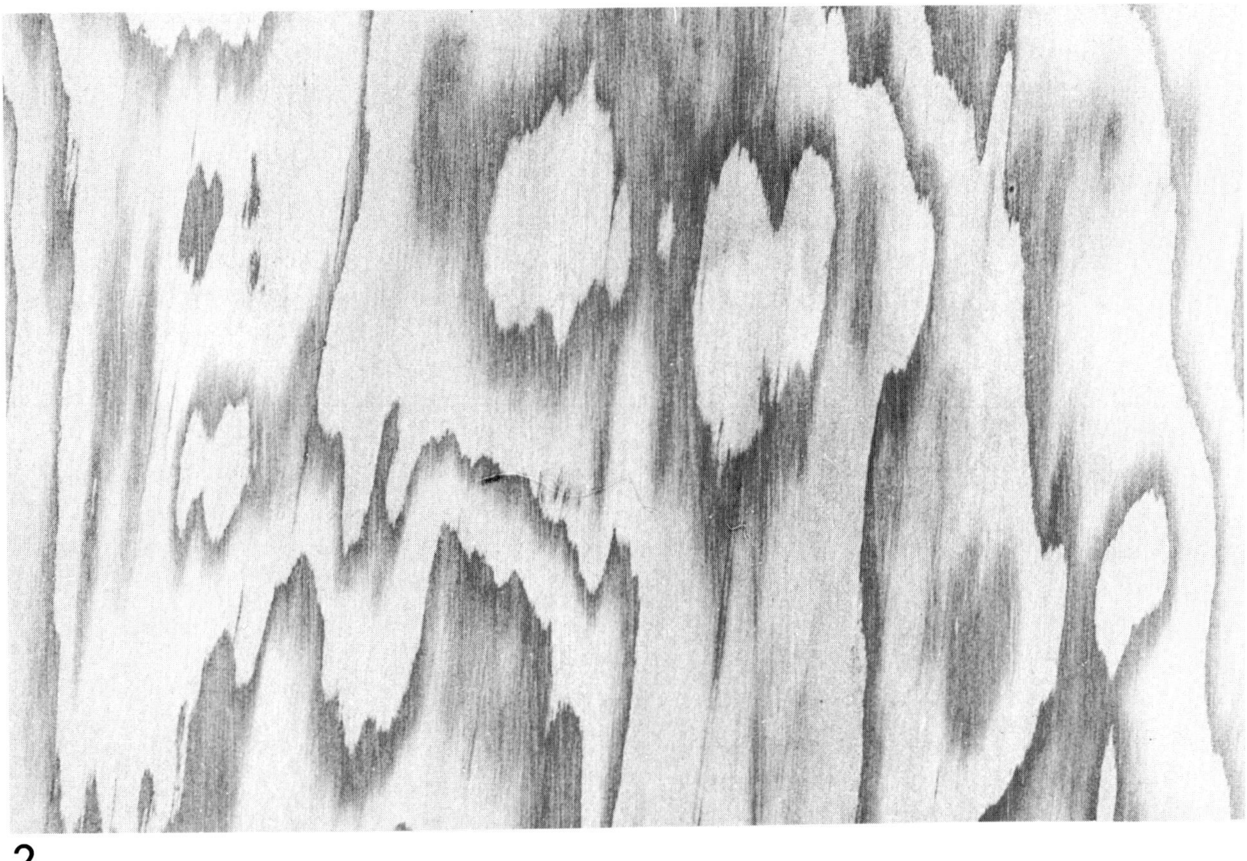

2.

Figure 109 Particleboard × 3

annual production was over one billion square feet, and rising. Presently, its chief use is for floor "underlayment." Conventional floor boards, or thick plywood sheets are nailed to the joists, and then a layer of particleboard is nailed on top. Its smooth and even surface forms an underlayment for wall to wall carpeting, or such coverings as linoleum or cork tiles. In furniture manufacture, particleboard is used as the core to which surface veneers are glued using modern resin adhesives. Decorative paneling is made in the same way. It would seem that the interesting pattern of the board itself would make it attractive as a wall covering, but there is a difference of opinion on its attractiveness.

The manufacture of particleboard (also called chipboard) starts with a machine that reduces the pieces of wood to smooth, thin chips or flakes that are approximately of uniform size. When planer shavings are used, part of this preparation has already been done. Green material of softwoods or the softer hardwoods is needed for best chipping. The chips must be carefully dried to a determined moisture content and then in a mixer are sprayed with a resin adhesive so that all the chip surfaces are covered. A thick mat of chips is then formed on a moving belt that carries the resin-coated chips to the hot presses. During pressing, the adhesive is set, but the boards are stacked and left for a few days to "cure." Finally they pass through a planer and sander. This short account gives no idea of the many variables that must be controlled all through the manufacturing process. They include the relation of chip moisture content to the dilution of the resin solution, mat thickness, time, temperature and force of pressing, and many others.

Paper, a Wondrous Material

The art of papermaking has made possible civilization as we know it, and the higher the standard of living, the more paper we use. The two tables below show (1) the enormous increase in the use of paper per person per year in the United States, and (2) the 1968 consumption of paper in several countries.

Use of Paper in Pounds per Capita

Increase in the U. S.		1968 use in several countries	
1800	3/4 lb.	United States	534
1850	7	Sweden	370
1900	58	Canada	368
1950	311	Switzerland	294
1968	534	Denmark	272
1975 est.	650	United Kingdom	270
		Mainland China	10

The first papermaker was a wasp, millions of years before man came on the scene. You have doubtless seen the beautiful rounded-conical structures made of layers of paper enclosing the brood combs. The wasp papermakers gnaw wood fibers, mix them with saliva and spread out the mixture in continuous sheets to build their nests. The first human papermaker was a Chinese, about 2,000 years ago, who managed to make sheets of paper from mixtures of bark and hemp fibers, and rags. It took a thousand years for the originally secret art of papermaking to reach Europe. The fascinating history of this art is one chapter of "Paper, the Fifth Wonder" (1) which gives a graphic account of how paper is made, including the chemical methods used.

Although the word "rags" sounds anything but elegant, the very finest papers were and still are made from them. Paper from linen and cotton fibers has marvelous strength and endurance. It is used for such things as paper money, bonds, documents, historical records, and the finest of writing papers. However, rags were always in short supply, and also the method of making paper in single sheets by hand was very time consuming. A fine screen, the size of the sheet to be made was dipped into a vat of fibers in water suspension, and then lifted to allow the water to drain away.

About one hundred years ago, a new and almost inexhaustible source of paper fibers was finally tapped—wood. It was found that by forcing a billet of wood against a grindstone upon which water was falling, a pulp was formed. This "groundwood" is made into newsprint and other papers. There were also soon developed several chemical processes which actually freed the fibers from each other by removing the lignin. With increased need for more paper, a machine with a moving endless wire screen belt was invented. The water-fiber mixture was fed to the screen, the water drained away, and the fibers felted together to form a continuous sheet of paper which passed through a series of heated drying cylinders before being wound in a roll at the end of the machine. Modern paper machines are enormous and when running full speed are awesome to watch. In a single 24 hour day, one machine may produce 700 miles of newsprint, or 1,100 miles of tissue (facial, or bathroom). For better grades of paper, the speed is less, but even so, tremendous quantities are produced. You can get an estimate by multiplying the per capita production by the population of the country.

Fig. 110 shows the wood fibers comprising a very thin sheet of paper. Paper is a controlled random arrangement of fibrous raw materials felted from a water suspension. Not only rags, and wood may be used but also other fibers both natural, and man made. Furthermore, a piece of paper is not just a piece of paper. In the beating machine, dyes to color it may be mixed with the fibers before they go to the paper machine. Other additives include clay, wax, starch, modified rosin from pine trees, and synthetic resins. These and other substances alone or in combination change greatly the qualities of the finished paper. It may be waterproofed, or given a highly reflective surface, made considerably stronger, or exhibit other qualities produced for the special use to which it will be put. If you look about you, the number of things obviously made of paper will surprise you. But there are many others, made for instance from molded pulp (egg cartons), which require careful inspection to discover their basic material. There are literally thousands of things made from wood pulp and paper, and since many of them are disposable

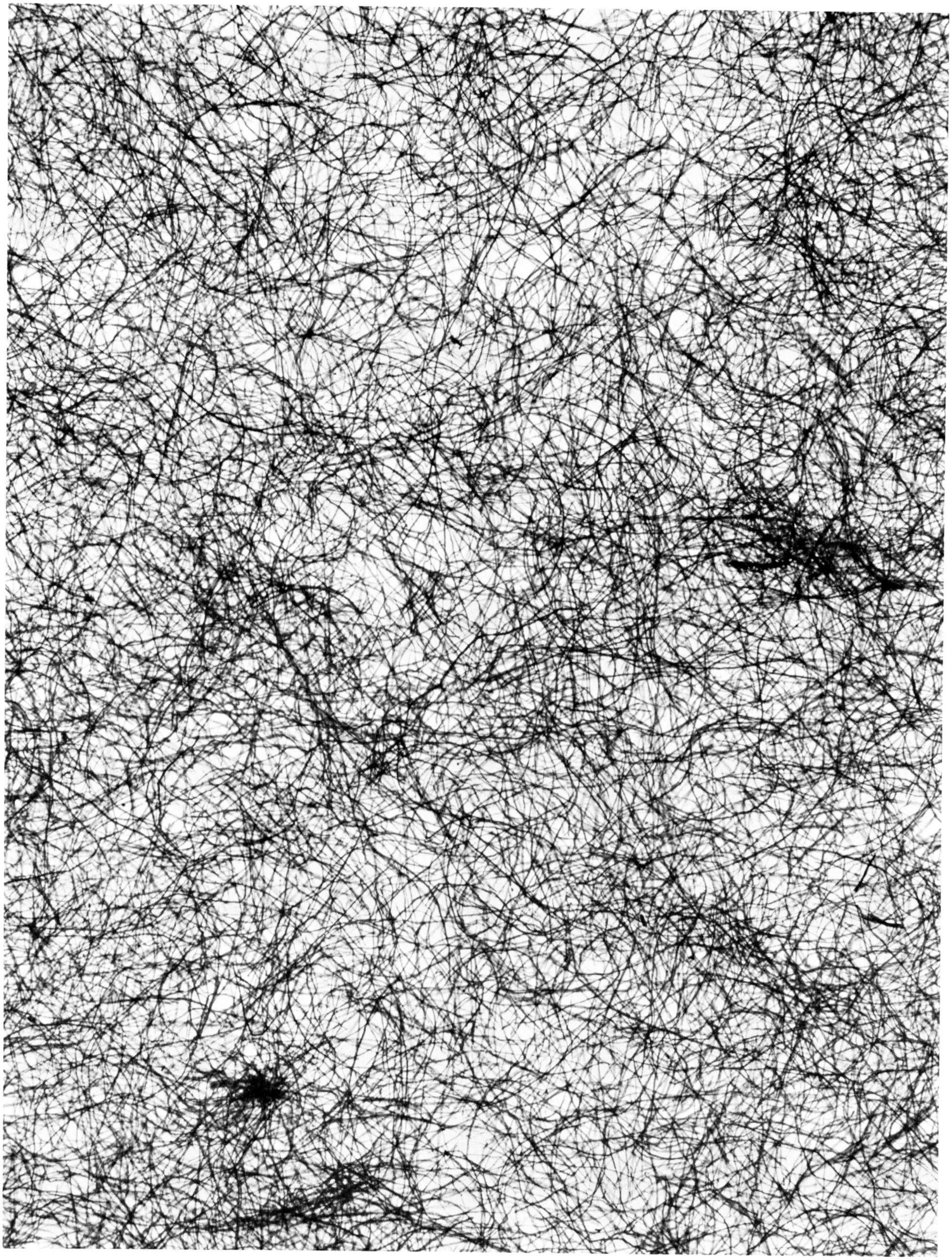

Figure 110 Southern pine wood fibers form a thin sheet of paper

after use (containers, etc.), we have the problem of what to do with them. About one third of the used paper is re-cycled to the machines, and this amount is increasing. Paper and paper products are now so important that without them our industrial civilization would grind to a halt.

Wood-plastic Composites

Various methods have been developed to alter the basic properties of wood with the hope that it might become less pervious to water, and consequent swelling. Such treatments also aim to increase the strength and surface hardness of wood. With the great interest in polymer (plastic*) chemistry, it is natural that these substances should be used to impregnate wood structure. Polymers are long chain molecules, each separate "link" of which is called the monomer. A string of plastic beads each of which is fastened to its neighbor by a flexible joint is often used as a "picture" of a polymer. One may start with a number of separate beads (the monomer) and by linking them together produce the long chain polymer.

Dry wood under vacuum can be directly impregnated with the liquid monomer. Early studies showed that polymerization to the solid state can be effected by radiation, but this involved an expensive radiation source, and special precautions and clearance from the Atomic Energy Commission. Dr. John A. Meyer, and his students and colleagues at State University of New York, College of Forestry at Syracuse, have found that it is only necessary to add a heat sensitive catalyst to the monomer before it enters the wood structure. Then, heating to a temperature of 150°F. produces a wood-plastic combination essentially the same as that developed by radiation. In either case, the wood-plastic material takes up water vapor from the surrounding air more slowly than does wood itself. Most physical properties of the new material are greatly enhanced. Compressive strength, hardness, resistance to abrasion, all are higher than before treatment; however, tensile strength remains the same. Thin sections of the treated wood show that the cell cavities are more or less filled with the plastic, and that some of it is even in the cell wall. The strength of the plastic itself is much less than that of the wood, therefore the greater strength of the combination must be due to the "bulking" effect of the plastic in the cell cavities. If you glue together some soda "straws", the resulting structure can be easily crushed, but a similar one filled with dirt or sand is much stronger because the filling keeps the thin walls of the structure from collapsing. This bulking effect makes the compressive strength of ice-filled wood greater in the same way. Wood-plastic material is obviously heavier than the untreated wood—often two or more times as much depending upon how much of the monomer is sucked into the cell cavities during treatment.

Exploration in this field of wood-plastic composites has scarcely begun. Only a few woods (mostly hardwoods with their large pores) have been used, and the number of plastics so far tried is small. Even so, great variation has already been discovered in the behavior of different woods in combination with a single plastic, or with different plastics. As an example, the monomer methyl methacrylate* upon polymerization shrinks nearly 25 percent in volume, and wood-plastic material made with it may have one third of its microscopic tubes unfilled. By contrast, t-butyl styrene shrinks only 7 percent, and the empty cavities are nearly filled with the plastic.

These materials can be worked with the usual wood working tools, but good tool steel, or carbide tips are recommended. When a power sander was used, the wood impregnated with methacrylate soon gummed up the sandpaper to the point where it was useless. The heat of sanding melted the plastic. This disadvantage was overcome by adding a cross-linking agent to the monomer. The polymerized plastic then had a higher melting point and, when sanded, decomposed before melting.

The wood using industry has only begun to

* The term "plastic" is not a precise one. It usually refers to man made polymers.

* The polymer is called Lucite or Plexiglas.

Figure 111 Wood sculpture using a strip of wood soaked in liquid ammonia. (by George F. Earle)

apply the results of research on wood-plastic composites. Presently, flooring is the most important use. The much greater surface hardness of the material, it is hoped, will resist wear and prolong the life of the floor. Time will tell. Numbers of small items and novelties are being made. These include the center portion of laminated archery bows, the front half of billiard cues, hair brush backs, and dynamite tamping sticks. These sticks are valued because they are heavy, and the tamping end resists wear better than un-impregnated wood. Undoubtedly these wood-plastic materials will have more widespread use as more is known about them, and this knowledge is spread throughout the industry.

The Bending of Wood

The production of permanent curves or bends in wood by steam treatment is a very ancient art. The steam plasticized piece of wood is bent around a form and clamped or otherwise held in place until dry. When the clamps are released, the piece retains its curve. As examples, consider snowshoe frames, ski tips, curved furniture parts, and many others.

Although it has been known for some time that ammonia, too, caused wood to swell and presumably become plastic, Dr. Conrad Schuerch of State University of New York, College of Forestry at Syracuse, in 1963, was the first to show some of the exciting things that can be done with ammonia-treated wood. An artist, Professor George F. Earle in the College's School of Landscape Architecture, saw the possibilities of ammonia-treated strips of wood in sculpture. He and his students were soon creating beautifully complex forms such as that in Fig. 111. Other artists have been quick to experiment with this new medium.

For others interested, it should be said at once that the *liquid* ammonia used in these experiments is not the "ammonia" sold in the supermarket. The latter is only a relatively mild water solution of the former. Ammonia is a gas at ordinary room temperatures. Liquid ammonia at 75°F in pressure tanks only remains a liquid as long as the gas above it is under a pressure of about 150 pounds per square inch, ten times that of the surrounding air. Liquid ammonia boils at −28°F, so in order to use it as a liquid, the container should be a few degrees below this point. A freezer chest with a loosely fitting cover, and capable of reaching extra low temperatures, was used in the first experiments. The wood strips were immersed in the liquid ammonia contained in a long shallow sheet iron pan, plastic or ceramic coated (ammonia reacts with copper, brass, or aluminum). Besides being explosive when mixed with air, ammonia gas is extremely corrosive to eyes, nose, lungs, and even the skin. The special freezer chest or any other container in which the liquid ammonia is exposed to the outer air must be installed outdoors, or under a hood with a fan to remove the fumes. The wood strips are immersed and withdrawn with tongs, and the hands and lower arms are protected by long rubber gloves. Over these a pair of heavy gloves is worn to insulate one from the intense cold of the withdrawn strip of ammonia-soaked wood. The time of immersion for $\frac{1}{16}$ inch material is about one half hour. One eighth inch thick pieces require four or five hours, and plasticity may be improved by leaving them overnight in the liquid ammonia. Because of the fumes, the hand bending must be done outdoors or under a hood. At first, the wooden strip is somewhat stiff but, as it approaches room temperature it becomes more plastic, and in working with it one is reminded of stiff leather. Quite sharp curves are possible, and a thin strip can even be tied into a knot. The ammonia quickly passes into the air, and forming must be completed within 15 to 30 minutes, depending upon the thickness of the piece. When thoroughly dried, the strip retains permanently its sculptured curves. If it is soaked in water, it tends to straighten, but when dried again, it returns to its previously formed shape. It is believed that the ammonia loosens the hydrogen bonds that cross-connect the long chain molecules of cellulose, and also loosens the bonds in the lignin molecules. When the piece of wood is bent, these macro-molecules can slide past each other. As the ammonia evaporates, connecting bonds are formed in the new position. This may account for the stability of the bent or curved strips. However, microscopic slip lines in the cell wall may also be involved.

Certain softwoods treated in this way tend to develop compression creases on the inner faces of the bends. Some ring-porous hardwoods with their relative weak zones of earlywood are also unsuitable. Perhaps the best woods are the diffuse porous ones of medium to heavy weight. These include beech, birch, maple and cherry, but probably many others can be worked with the same ease.

Following the publication of the liquid ammonia method, other research workers found that equally good results could be obtained faster by using ammonia gas in an iron or steel pressure tank at room temperature. After placing the wood strips in the bottom of the long tank, it was sealed and air withdrawn. Then from a pressure cylinder of liquid ammonia, the gas at the top was allowed to pass into the treatment tank. The gas developed its own pressure which at 75°F was about 150 pounds per square inch. After the treatment period of about an hour (the time depends upon the thickness and moisture content of the wood) the pressure cylinder valve was closed, and the tank vented to the outdoors. Both liquid and gaseous methods are patented.* In summary, either method is safe when supervised by a chemist or chemical engineer. *No one* should attempt to try it without such supervision. Serious damage to the human body can occur when either liquid or gaseous ammonia is used without proper equipment and a knowledge of its corrosive properties, and the possibility of an explosion.

* Both the liquid and gaseous method patents are administered by the Research Corporation, 405 Lexington Avenue, New York, N. Y. 10017.

Apologia

Because this short introduction to wood structure and properties is not intended to cover the whole field of wood technology, perhaps no apology is necessary. However, I should like to mention some of the other important things that the reader will find in "Textbook of Wood Technology" (8). These include variability of wood structure within the same tree, and between different trees of the same kind. Although defects such as knots, and reaction wood have been described, some others not included are brashness, pitch and bark pockets, insect damage, and that caused by improper drying and machining. The natural weathering of wood over long periods of time had also been omitted.

It is hoped that this highly pictorial approach and short text will intrigue many of you to explore further "WOOD—Masterpiece of Nature."

Appendix

How to Separate Wood Fibers*

This chemical process called maceration dissolves the middle lamella and liberates the woody cells so that they either float free from each other, or do so upon slight shaking of the container. Mild maceration and shaking often produce extremely interesting cell aggregates. You may see several fibers, a vessel or two, and a plate of ray cells all still attached to each other.

Of course chemical pulping methods produce maceration, but are not available to someone who wants to look at only a few hundred fibers.

To prepare wood fibers, first saw off small pieces of wood making them about ¾ inches long, and then split them into match stick size. You need enough to fill a 40 ml (c.c.) test tube about ¼ full. Now, in a beaker or other container, boil the sticks in water until they sink or become waterlogged. Pour into the bottom of the dry test tube between one and two grams (the exact amount is not critical) of the dry chemical potassium chlorate. Drop the boiled sticks on top of the chemical, and place the tube in a test tube rack or other holder. Using great care, pour in concentrated nitric acid until it a little more than covers the wood. Nitric acid causes severe burns on the skin. If any acid touches the skin, wash it off at once with plenty of water. Any spillage must be thoroughly diluted at once and washed down the sink. Nitric acid and potassium chlorate are powerful oxidizing agents, and immediately begin to react with each other. Several poisonous gases are produced certain of which, including chlorine in solution, remove the lignin. At least one of the gases is explosive, but the amounts produced here are quite small, and are not dangerous. Lacking a chemical laboratory hood, place the test tube rack with its one or some tubes outdoors, conveniently perhaps on a window sill. There it is to stay for several days to a week. Each day look at the little wood sticks and notice that they keep getting paler and paler. When they are pure white, bring the tube in, and slowly pour in water to dilute the acid. Pour the diluted acid into a beaker of water to dilute the acid further before it goes down the sink. Wash the cellulose sticks with three changes of water. All this has to be done carefully or you may lose too much of the material down the sink. If the chemical reactions have worked as desired, the sticks after the third washing will be just about ready to fall apart. Fill the tube about halfway with water, place your thumb over the end of the tube and shake. Some of the fiber mass can now be poured out upon a large microscope slide or in to a small glass dish for observation. Hopefully your microscope is equipped to use polarized light. Beautiful rainbow colors appear and disappear in the fibers as the circular microscope stage is rotated.

It is possible, especially in cold weather if the tube is left outdoors, that the reaction will progress too slowly. If so, warm the tube *gently* until bubbles of reddish brown gas emerge and begin to fill the upper part of it. Then return the tube to the hood or outdoors. Any sudden or too much heat will produce a miniature volcano, a very dangerous effect.

Because of the precautions given, do not fear to try the method. Anyone having taken a course in beginning chemistry would be familiar with the nature of the two reagents used. The precautions are for those who have not been so privileged.

---

* Another method using strong household bleach, but requiring an oven is given in "Classroom Demonstrations of Wood Properties" (16).

Relative Resistance of Heartwood to Decay (Adapted from U. S. Forest Products Lab., Tech. Note 229, 1961)

High	Intermediate	Low
Baldcypress (old growth)	Baldcypress (young growth)	Alder, red
Catalpa	Douglas-fir	Ashes
Cedars	Honeylocust	Aspens
Chestnut	Larch, western	Basswood
Cypress, Arizona	Pine, eastern white	Beech
Junipers	Pines, southern	Birches
Locust, black	Tamarack	Buckeye
Mulberry, red		Butternut
Oaks, white (most)		Cottonwoods and other poplars
Osage-orange		Elms and hackberry
Redwood		Hemlocks
Sassafras		Hickories
Walnut, black		Maples
Yew, Pacific		Oaks, red
		Pines (most others)
		Spruces
		Sweetgum
		Sycamore
		Willows
		Yellow-poplar

These woods are listed approximately alphabetically, and are definitely not in order of decay resistance in each group. Also because of the great variability of wood itself and the conditions under which it might be used, to say nothing of the selectivity of certain fungi in attacking some woods but not others, these listings must be considered only as estimates. However, if this is understood the table is still useful, especially in comparing woods with high resistance with those of low or almost no resistance.

Selected References

1. Ainsworth, J. H. "Paper, the Fifth Wonder," Thomas Printing and Publishing Co. Ltd., Kaukauna, Wisconsin. 1967.

2. Albion, R. G., "Forests and Sea Power," Harvard Univ. Press, Cambridge, Massachusetts. 1926.

3. Browne, F. L. Theories of the Combustion of Wood and its Control. Report 2136. December 1958, information reaffirmed 1963. U. S. Forest Products Laboratory, Madison, Wisconsin.

4. Côté, W. A., Jr., (Editor), "Cellular Ultrastructure of Woody Plants," Syracuse Univ. Press, Syracuse, New York. 1965.

5. Kollmann, F. F. P. and Côté, W. A., Jr., "Principles of Wood Science and Technology," Vol. 1, Springer-Verlag, New York. 1968.

6. Mark, H. F., "Giant Molecules," Time-Life Books, New York, New York. 1968.

7. National Forest Products Assoc., 1619 Massachusetts Ave., Washington, D. C. 20036. The Challenge of the Forest (free booklet).

7a. National Lumber Manufacturers Assoc., Washington, D. C. Lumber and Wood Products Literature.

8. Panshin, A. J., de Zeeuw, C. and Brown, H. P., "Textbook of Wood Technology," Vol. 1, 3rd Ed., McGraw-Hill Book Co., New York. 1970.

9. Panshin, A. J., Harrar, E. S., Baker, W. J. and Proctor, P. B., "Forest Products," McGraw-Hill Book Co., New York. 1950.

10. Shigo, A. L. and E. H. Larson., A Photo Guide to the patterns of discoloration and decay in living northern hardwood trees, U. S. Forest Service Research Paper NE-127. 1969.

11. State Univ. of New York College of Forestry, Syracuse, New York 13210. Careers in Wood Products Engineering (free booklet).

12. Tiemann, H. D., "Wood Technology," 3rd Ed. Pitman Publ. Corp., New York. 1951.

13. U. S. Forest Products Lab., "Wood Handbook," Agricultural Handbook No. 72. 1955. Supt. of Documents, Washington, D. C. 20402. Price $2.25

14. U. S. Forest Service, Challenge in Wood Research, Misc. Publ. No. 1054. Washington, D. C. 1967.

15. U. S. Forest Service. "Wood . . . Colors and Kinds," Agricultural Handbook No. 101. 1956. Supt. of Documents, Washington, D. C. 20402. Price $.50

16. —— Classroom Demonstrations of Wood Properties, PA-900. 1969. Supt. of Documents, Washington, D.C. 20402. Price $.60

16a. —— Products of American Forests, Misc. Publ. 861, 1969. Supt. of Documents. Price $.50

17. Westing, A. H., Formation and Function of Compression Wood in Gymnosperms, Botan. Rev. 34:51-78. 1968.

18. Wise, L. E. and E. C. Jahn, "Wood Chemistry," 2nd Ed. Reinhold Publ. Co., New York. 1952.

Wood Sample Collection. 20 different species. $4.50 per set. Turtox Cat. No. 25D7261. General Biological, Inc., 8200 South Hoyne Ave., Chicago, Illinois 60620.

WOOD—Masterpiece of Creation. A 28 min. 16 mm film produced by W. M. Harlow for the New York State College of Forestry. Syracuse, N. Y. 13210. New York State users apply to the Film Library. Out of state requests for rental or preview should be sent to the International Film Bureau, 332 S. Michigan Ave. Chicago, Ill. 60604.

INDEX*

* Boldface numbers refer to species illustrations and uses.